Lecture Notes in Mathematics

Edited by A. Dold and B. Eckmann

895

Jonathan A. Hillman

Alexander Ideals of Links

Springer-Verlag
Berlin Heidelberg New York 1981

Author

Jonathan A. Hillman
Department of Mathematics, University of Texas
Austin, TX 78712, USA

AMS Subject Classifications (1980): 13C99, 57M25, 57Q45

ISBN 3-540-11168-9 Springer-Verlag Berlin Heidelberg New York
ISBN 0-387-11168-9 Springer-Verlag New York Heidelberg Berlin

Printing and binding: Beltz Offsetdruck, Hemsbach/Bergstr.
2141/3140-543210

PREFACE

The characteristic polynomial of a linear map is one of the most basic of mathematical objects, and under the guise of the Alexander polynomial has been much studied by knot theorists. The rational homology of the infinite cyclic cover of a knot complement is indeed determined by a family of such polynomials. The finer structure of the integral homology, or the homology of covers of a link complement (corresponding to a set of commuting linear maps) is reflected in the Alexander ideals. These notes are intended to survey what is presently known about the Alexander ideals of classical links, and where possible to give "coordinate free" arguments, avoiding explicit presentations and using only the general machinery of commutative and homological algebra. This has been done to clarify the concepts; in computing examples it is convenient to use Wirtinger presentations and the free differential calculus, Seifert surfaces or surgery descriptions of links. (The avoidance of techniques peculiar to the fundamental group or to 3-dimensional topology means also that these arguments may apply to links in higher dimensions, but little is said on this topic after Chapter II.)

This work grew out of part of my 1978 A.N.U. Ph.D. thesis. However although most of the proofs are mine, a number of the results (mostly in Chapters I, IV, VII and VIII) are due to others. Some of the latter results have been quoted without proof, as the only proofs known to me are very different in character from the rest of these notes.

I would like to acknowledge the support of a Commonwealth Postgraduate Research award at the Australian National University while writing my thesis, and of a Science Research Council grant at the University of Durham while preparing these notes. I would also like to thank Professors Levine, Murasugi, Sato and Traldi for sending some of their (as yet) unpublished notes to me. Finally I would like to thank Mrs. J. Gibson and Mrs. S. Nesbitt for the care with which they have prepared the typescript.

CONTENTS

PRELIMINARIES

In these notes we shall generally follow the usage of Bourbaki [13] for commutative algebra, Crowell and Fox [43] for combinatorial group theory, and Rourke and Sanderson [159] for geometric topology. The book of Magnus, Karrass and Solitar [123] is a more comprehensive reference for combinatorial group theory, while the books of Hempel [69], Rolfsen [157] and Spanier [177] are useful for other aspects of topology.

All manifolds and maps between them shall be assumed PL unless otherwise stated. The expression $A \approx B$ means that the objects A and B are isomorphic in some category appropriate to the context. When there is a canonical isomorphism, or after a particular isomorphism has been chosen, we shall write $A = B$. (For instance the fundamental group of a circle is isomorphic to the additive group of the integers $\mathbb{Z}$, but there are two possible isomorphisms, and choosing one corresponds to choosing an orientation for the circle).

Qualifications and subscripts shall often be omitted, when there is no risk of ambiguity. In particular "μ-component n-link" may be abbreviated to "link", and the symbols Λ_μ, $X(L)$, $G(L)$ may appear as Λ, X and G.

CHAPTER I LINKS AND LINK GROUPS

This chapter is principally a resumé of standard definitions and theorems, without proofs. Although our main concern is with the classical case, we have framed our definitions so as to apply also in higher dimensions. We begin with definitions of links and of the most important equivalence relations between them. Next we consider link groups and homology boundary links. There follows a section on the equivariant homology of covering spaces of link exteriors, and we conclude with some comments on the construction of such covering spaces.

Let μ and n be positive integers. If X is a topological space, let μX be the space $X \times \{1, \ldots, \mu\}$, the disjoint union of μ copies of X. Let $D^n = \{\langle x_1, \ldots, x_n\rangle \text{ in } \mathbb{R}^n \mid \sum_{1 \leq i \leq n} x_i^2 \leq 1\}$ be the n-disc and let $S^n = \partial D^{n+1}$ be the n-sphere. The standard orientation of $\mathbb{R}^n$ [159 ; page 44] induces an orientation of D^n, and hence of S^{n-1} by the convention that the boundary of an oriented manifold be oriented compatibly with taking the inward normal last (cf [159 ; page 45]).

Definition A μ-component n-link is an embedding $L: \mu S^n \longrightarrow S^{n+2}$. The ith component of L is the n-knot (1-component n-link) $L_i = L \mid S^n \times \{i\}$. A link type is an equivalence class of links under the relation of being ambient isotopic.

Notice that with this definition, and with the above conventions on the orientation of the spheres, all links are oriented. A 1-link is locally flat (essentially because there are no knotted embeddings of S^0 in S^2), but embeddings of higher dimensional manifolds in codimension 2 need not be locally flat [161 ; page 59].

Definition An I-equivalence between two embeddings $f,g:A \to B$ is an embedding $F:A \times [0,1] \longrightarrow B \times [0,1]$ such that $F|A \times \{0\} = f$, $F|A \times \{1\} = g$ and $F^{-1}(B \times \{0,1\}) = A \times \{0,1\}$.

In this definition we do not assume that the data are PL (so here an embedding is a 1-1 map inducing an homeomorphism with its image). Recent results of Giffen suggest that wild I-equivalences have a rôle even in the context of PL links [54a]. Clearly isotopic embeddings are I-equivalent.

A locally flat isotopy is an ambient isotopy [159 ; page 58], but even an isotopy of 1-links need not be locally flat. For instance any knot is isotopic to the unknot, but no such isotopy of a non trivial knot can be ambient. However a theorem of Rolfsen shows that the situation for links is no more complicated.

Definition Two μ-component n-links L and L' are locally isotopic if there is an embedding $j:D^{n+2} \longrightarrow S^{n+2}$ such that $D = L^{-1}(j(D^{n+2}))$ is an n-disc in one component of μS^n and such that $L|\mu S^n - D = L'|\mu S^n - D$.

Theorem (Rolfsen [153]) Two n-links L and L' are isotopic if and only if L' may be obtained from L by a finite sequence of local isotopies and an ambient isotopy.

In other words L and L' are isotopic if and only if L' may be obtained from L by successively suppressing or inserting small knots in one component at a time.

Definition A concordance between two μ-component n-links L and L' is a locally flat I-equivalence $\mathcal{L}$ between L and L'. A link is null concordant (or slice) if it is concordant to the trivial link.

A link L is a slice link if and only if it extends to a locally flat embedding $C:\mu D^{n+1} \longrightarrow D^{n+3}$ such that $C^{-1}(S^{n+2}) = \mu S^n$.

<u>Definition</u> Two μ-component n-links L and L' are <u>link-homotopic</u> if there is a map $H:\mu S^n \times [0,1] \longrightarrow S^{n+2}$ such that $H|\mu S^n \times \{0\} = L$, $H|\mu S^n \times \{1\} = L'$ and $H(S^n \times \{t\} \times \{i\}) \cap H(S^n \times \{t\} \times \{j\}) = \emptyset$ for all t in $[0,1]$ and for all $1 \leq i \neq j \leq \mu$.

In other words a link-homotopy is a homotopy of the maps L and L' such that at no time do the images of distinct components of μS^n intersect (although self intersections of components are allowed). Milnor [129] has given a thorough investigation of homotopy of 1-links. Giffen [55] and Goldsmith [57] have recently shown that concordant 1-links are link-homotopic. (Giffen [54a] has also shown that I-equivalent links need not be PL I-equivalent). For other results on isotopy of links and related equivalence relations see [26, 82, 111, 130, 154, 155,175].

The link group

The basic algebraic invariant of a link is the fundamental group of its complement, and most of these notes are concerned with the structure of metabelian quotients of the groups of 1-links.

<u>Definition</u> The <u>exterior</u> of a μ-component n-link L is $X(L) = S^{n+2} - N$, where N is an open regular neighbourhood of the image of L. The <u>group</u> of the link L is G(L), the fundamental group of X(L).

The exterior of L is a deformation retract of $S^{n+2} - L$, the complement of L, and is a compact connected PL (n+2)-manifold with μ boundary components. By Alexander duality $H_1(X(L);\mathbb{Z}) \cong \mathbb{Z}^\mu$, $H_i(X(L);\mathbb{Z}) = 0$

for $1 < i < n+1$ and $H_{n+1}(X(L);\mathbb{Z}) \approx \mathbb{Z}^{\mu-1}$. We shall assume henceforth that all links are locally flat. Then $\partial X(L) = \mu S^n \times S^1$. A meridianal curve for the ith component of L is an oriented curve in the boundary of X(L) which bounds a disc in $S^{n+2} - X(L_i)$ having algebraic intersection +1 with L_i. The image of such a curve in the link group G is well defined up to conjugation, and any element of G in this conjugacy class is called an ith meridian. The images of the meridians in the abelianization $G/G' = H_1(X:\mathbb{Z})$ are well defined and freely generate it, inducing an isomorphism with $\mathbb{Z}^\mu$.

An application of van Kampen's theorem shows that G is the normal closure of the set of its meridians. (The normal closure of a subset S of a group is the smallest normal subgroup containing S, and shall be denoted $\ll S \gg$). Thus the group of a μ-component n-link is a finitely presentable group G which is normally generated by μ elements, with abelianization $\mathbb{Z}^\mu$ and, if $n \geq 2$, with $H_2(G;\mathbb{Z}) = 0$ (since by Hopf's theorem $H_2(G;\mathbb{Z})$ is the cokernel of the Hurewicz homomorphism $\pi_2(X) \longrightarrow H_2(X;\mathbb{Z})$ [83]). Conversely Kervaire has shown that if $n \geq 3$ these four conditions characterize the group of a μ-component n-link [96]. If $n = 2$ these conditions are neccessary but not sufficient, even for $\mu = 1$ [71]; if the last condition is replaced by "the group has a presentation of deficiency μ" then Kervaire showed also that it is the group of a link in some homotopy 4-sphere, but this stronger condition is not neccessary.

The case of 1-links with $\mu > 1$ is quite different. For then $H_2(G;\mathbb{Z}) \approx \mathbb{Z}^{\mu-1}$ unless $\pi_2(X) \neq 0$, in which case by the Sphere Theorem [147] the link is splittable. (An n-link L is splittable if there is an (n+1)-sphere $S^{n+1} \subsetneq S^{n+2} - L$ such that L meets each complementary ball, that is, each component of $S^{n+2} - S^{n+1}$). This is related to the presence of longitudes, non trivial elements of the group commuting with meridians.

Let L be a μ-component 1-link. An ith longitudinal curve for L is a closed curve in the boundary of X(L) which is parallel to L_i (and so in particular intersects an ith meridianal curve in just one point), and which is null homologous in $X(L_i)$. The ith meridian and i^{th} longitude of L, the images of such curves in G(L), are well defined up to simultaneous conjugation. If X(L) has been given a basepoint *, then representatives of the conjugacy classes of the meridians and longitudes in $\pi_1(X(L),*) \approx G(L)$ may be determined on choosing paths joining each component of the boundary to the base point. The linking number ℓ_{ij} of the ith component of L with the jth is the image of an ith longitude of L in $H_1(X(L_j);\mathbb{Z}) = \mathbb{Z}$; it is not hard to show that $\ell_{ij} = \ell_{ji}$. (Notice that $\ell_{ii} = 0$).

When chosen as above, the ith longitude and i^{th} meridian commute, since they both come from the fundamental group of the ith boundary component, which is a torus. In the case of higher dimensional links there is no analogue of longitude in the link group, because spheres of dimension greater than or equal to 2 are simply connected, while in knot theory the longitudes are often overlooked, as for 1-knots they always lie in the second commutator subgroup G" (See below). The presence of the longitudes gives the study of classical links and their groups much of its special character.

If the ith longitude is equal to 1 in G(L), then L_i extends to an embedding of a disc disjoint from the other components of L, by the Loop Theorem [147]. A link is trivial if all the longitudes equal 1.

Theorem 1 A 1-link L is trivial if and only if G(L) is free.

Proof (Note that the rank of G(L) must equal the number of components of L). Since a free group contains no noncyclic abelian subgroups [123; page 42], the ith longitude and i^{th} meridian must lie in a common cyclic group. On considering the images in $H_1(X(L_i);\mathbb{Z}) = \mathbb{Z}$, we conclude that the ith longitude must be null homotopic. Hence using the Loop Theorem inductively we see that the link is trivial. The argument in the other direction is immediate. //

This result may be restated as "An n-link is trivial if and only if the homotopy groups $\pi_j(X)$ are those of a trivial link for $j \leq [\frac{n+1}{2}]$" and in this form remains true for n-knots whenever $n \geq 3$. (Of course the proofs are quite different. See Levine [114] for $n \geq 4$ and for $n = 3$ see Shaneson [169] in conjunction with Milnor duality [132]). However it is false for all $\mu \geq 2$ and $n \geq 2$, as was first shown by Poenaru [149]. (See also Sumners [181] and the remarks following Theorem II.6 below).

Definition A μ-component n-link L is a boundary link if there is an embedding $P: W = \cup W_i \longrightarrow S^{n+2}$ of μ disjoint orientable (n+1)-manifolds each with a single boundary component, such that $L = P|\partial W$.

All knots are boundary links, and conversely many arguments and results about knots proved by means of such "Seifert surfaces" carry over readily to arbitrary boundary links.

Theorem (Smythe [174]) A μ-component 1-link is a boundary link if and only if there is a map of G(L) onto F(μ), the free group of rank μ, carrying some set of meridians to a basis of F(μ).

Gutiérrez extended Smythe's theorem to n-links and characterized the trivial n-links for $n \geq 4$ as the boundary links whose complement has the correct homology groups $\pi_j(X)$ for $j \leq [\frac{n+1}{2}]$ [61]. The splitting theorem of Cappell shows that this is also the correct criterion for n = 3 [22]. (Little is known about the case n = 2, even for knots. See Swarup [187]).

Definition A μ-component link L is an homology boundary link if there is an epimorphism $G(L) \longrightarrow F(\mu)$.

Note that there is no assumption on the meridians. The kernel of such an epimorphism is necessarily $G_\omega = \bigcap_{n \geq 0} G_n$, the intersection of the intersection of the terms of the lower central series of G. (See below in this section). Smythe showed also that a 1-link L is an homology boundary link if and only if there are μ disjoint orientable surfaces U_i in X(L) with $\partial U_i \subset \partial X(L)$ and such that ∂U_i is homologous to the ith longitude in $\partial X(L)$. (Such surfaces shall be referred to as "singular Seifert surfaces"). For an homology boundary link, the longitudes lie in G_ω, since a free group contains no noncyclic abelian subgroups. For a boundary link, they lie in $(G_\omega)'$, since they bound surfaces which lift to the maximal free cover of the link complement. (See also the next section).

Any 1-link is ambient isotopic to a link L with image lying strictly above the hyperplane $\mathbb{R}^2 \times 0$ in $\mathbb{R}^3 = S^3-\{\infty\}$ and for which the composition poL with the projection $p:\mathbb{R}^3 \longrightarrow \mathbb{R}^2$ is local embedding with finitely many double points. Given such a link, a presentation for the link group (the Wirtinger presentation) may be found in the following way. For each component of the link minus the lower member of each double point pair assign a generator. (This will correspond to a loop coming in on a straight line from ∞, going once around this component, and returning to ∞). For the double point corresponding to the arc x crossing over the point separating arcs y and z, there is a relation $xyx^{-1} = z$, where the arcs are oriented as in Figure 1.

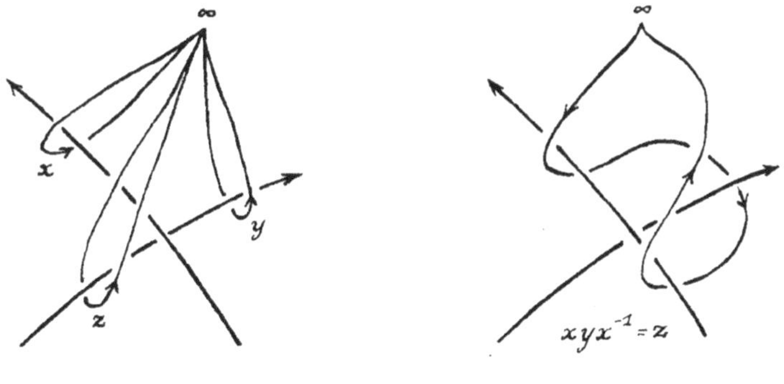

Figure 1

This gives a presentation of deficiency 0 for G(L), of the form

$$\{x_{ij},\ 1 \leq j \leq j(i),\ 1 \leq i \leq \mu \mid u_{ij}\, x_{ij}\, u_{ij}^{-1} = x_{ij+1},\ 1 \leq j \leq j(i),\ 1 \leq i\}$$

(where $u_{ij} = x_{pq}^{\pm 1}$ for some p, q and $x_{ij(i)+1} = x_{i1}$). It is not hard to show that one of these relations is redundant. (See Crowell and Fox [43; pages 72-86] for details). Thus a 1-link group has a presentation of deficiency 1. For a knot group this is clearly best possible .

Theorem 2 The group G of a link L has a presentation of deficiency greater than 1 if and only if L is splittable.

Proof If G has a presentation with a generators and b relations, then G is the fundamental group of a 2-dimensional cell complex Z with 1 0-cell, a 1-cells, and b 2-cells, so rank $H_2(G;\mathbb{Z}) \leq$ rank $H_2(Z;\mathbb{Z})$ = rank $H_1(Z;\mathbb{Z})+b-a$. Therefore if $a - b > 1$ then rank $H_2(G;\mathbb{Z}) \leq \mu - 2 <$ rank $H_2(X(L);\mathbb{Z})$ so $\pi_2(X(L)) \neq 0$, and so by the Sphere Theorem X(L) contains an embedded essential 2-sphere which must split L. The argument in the other direction is immediate, since the group of a splittable link is the free product of 2 link groups. //

As in [62] the group G(L) can be given a "preabelian" presentation [123 ; page 149] of the form

$$\{x_i, y_{ij}, 2 \leq j \leq j(i), 1 \leq i \leq \mu \mid [v_{ij}, x_i]y_{ij}, [w_i, x_i], 2 \leq j \leq j(i), 1 \leq i \leq \mu\}$$

where the v_{ij} and w_i are words in the generators x_i and y_{ij} and where the word w_i represents an ith longitude in G(L). (Notice that the generator x_i here, and all the generators x_{ij} for $1 \leq j \leq j(i)$ in the Wirtinger presentation are representatives of the ith meridians).

Theorem (Milnor [130]) The nilpotent quotient G/G_n of a link group G has a presentation of the form

$\{x_i, 1 \leq i \leq \mu \mid [w_i^{(n)}, x_i], 1 \leq i \leq \mu, F(\mu)_n\}$ where $w_i^{(n)}$ is a word in the generators representing the image of the ith longitude.

These nilpotent quotients are of particular interest because of the following result.

Theorem (Stallings [178]) If $f:H \longrightarrow K$ is an homomorphism inducing an isomorphism on first homology (abelianization) and an epimorphism on second homology (with coefficients in the trivial module $\mathbb{Z}$) then f induces isomorphisms on all the nilpotent quotients $f_n:H/H_n \xrightarrow{\simeq} K/K_n$. Consequently, if $\mathscr{L}$ is an I-equivalence of two links L_o and L_1, then the natural maps $G(L_o)/G(L_o)_n \xrightarrow{\simeq} G(\mathscr{L})/G(\mathscr{L})_n$ are isomorphisms, and so the nilpotent quotients of the link group are invariant under I-equivalence.

Here $G(\mathscr{L})$ denotes $\pi_1(S^3 \times [0,1] - \mathscr{L})$. If L is an homology boundary link, the epimorphism $G(L) \longrightarrow F(\mu)$ satisfies the hypotheses of Stallings' theorem, and so $G/G_n \xrightarrow{\simeq} F(\mu)/F(\mu)_n$ for all integers $n \geq 1$. Hence $G/G_\omega \twoheadrightarrow F(\mu)/F(\mu)_\omega = F(\mu)$, since free groups are residually nilpotent [145 ; page 112]. If G is the group of a higher dimensional link then the inclusion of a set of meridians induces a map $F(\mu) \longrightarrow G$ which also satisfies the hypotheses of Stallings' theorem, so again $G/G_n \approx F(\mu)/F(\mu)_n$. In this case however we cannot assume that the map $F(\mu) \longrightarrow G/G_\omega$ is onto, although it is 1-1.

Equivariant (co)homology

Let L be a μ-component n-link and let $p:X' \to X$ be the maximal abelian cover of the exterior of L. On choosing fixed lifts of the cells of X to X' we obtain a finite free basis for C_*, the cellular chain complex of X', as a $\mathbb{Z}[G/G']$-module. The isomorphism determined by the meridians enables us to identify $\mathbb{Z}[G/G']$ with $\Lambda_\mu = \mathbb{Z}[\mathbb{Z}^\mu] = \mathbb{Z}[t_1,t_1^{-1}, \ldots t_\mu,t_\mu^{-1}]$, the ring of integral Laurent polynomials in μ variables. This ring is a regular noetherian domain of dimension $\mu + 1$, and in particular is factorial. As a group ring Λ_μ has a natural involution, denoted by an overbar, sending each t_i to $\bar{t}_i = t_i^{-1}$, and augmentation $\varepsilon:\Lambda \to \mathbb{Z}$, which sends each t_i to 1.

Let $H_p(X;\Lambda)$ denote the Λ-module $H_p(C_*)$, which is just $H_p(X';\mathbb{Z})$ considered as a Λ-module via the covering transformations, and let $H^p(X;\Lambda)$ denote the pth cohomology module of the dual complex $Hom_\Lambda(C_*,\Lambda)$. (This may be regarded as the pth cohomology with compact supports of X'). Since C_* is a finite free complex and Λ is noetherian, all these homology and cohomology modules are finitely generated. The cohomology modules may be related to the homology modules by the Universal Coefficient spectral sequence [158 ; page 347] :

$$Ext^q_\Lambda(H_p(X;\Lambda),\Lambda) \Longrightarrow H^{p+q}(X;\Lambda)$$

There is also a Cartan-Leray spectral sequence [158 ; page 345] :

$$Tor^\Lambda_p(H_q(X;\Lambda),\mathbb{Z}) \Longrightarrow H_{p+q}(X;\mathbb{Z})$$

relating the equivariant homology to the homology of the base. (Note that $\mathbb{Z}$ is a Λ-module via the augmentation map. We shall not need the corresponding Cartan-Leray spectral sequence for cohomology). If $q:(\tilde{Y},\tilde{Z}) \longrightarrow (Y,Z)$ is any regular cover of a simplicial pair, there are similar equivariant (co)homology modules and spectral sequences.

Now since X is a compact PL (n+2)-manifold with boundary, there are Poincaré duality isomorphisms [131]:

$$\overline{H}^p(X;\Lambda) \cong H_{n+2-p}(X,\partial X;\Lambda)$$

given by cap product with the orientation class in $H_{n+2}(X,\partial X;\mathbb{Z})$. (Here if A is a Λ-module, $\bar{A}$ denotes the conjugate Λ-module, with the same underlying abelian group but with Λ-action given by $\langle\lambda,a\rangle \longmapsto \bar{\lambda}.a$ for all a in A, λ in Λ). This map may be interpreted geometrically in terms of intersections of dual cells in X' as was done by Blanchfield [11]. (See Chapter IX).

Other covers of a link exterior may be treated in the same way. In particular if L is an homology boundary link we may consider the maximal free cover $X^{\omega} \to X$. In this case the coefficient ring $\mathbb{Z}[G/G_{\omega}] \approx \mathbb{Z}[F(\mu)]$ is coherent [200], so all the equivariant (co)homology modules are finitely presentable, and of global dimension 2 , so the spectral sequences are fairly tractable. However since this group ring is not commutative, the distinction between left and right modules must be observed. (Taking the dual or the conjugate converts left to right and vice versa). These facts have been applied to boundary links by Sato [162, 164]. If L is an homology boundary 2-link, the Universal Coefficient spectral sequence together with Poincaré duality gives an isomorphism

$$\overline{e^2(G_{\omega}/G_{\omega}')} \approx e^2\, e^2(G_{\omega}/G_{\omega}')$$

(where $e^q(M) = \mathrm{Ext}^q_{\mathbb{Z}[G/G_{\omega}]}(M, \mathbb{Z}[G/G_{\omega}])$). This isomorphism can probably be used to show that there is a 3-link group which is not a 2-link group, although the groups of the component knots are 2-knot groups. (In the knot theoretic case such an isomorphism was used by Levine [120] to deduce that the p-local Alexander invariants of a 2-knot are symmetric and hence that not every high dimensional knot group is a 2-knot group).

Covering spaces for link exteriors may often be constructed by splitting along Seifert surfaces. This technique in conjunction with the Mayer-Vietoris sequence leads to presentations of the equivariant homology modules. In the case of the first homology these presentations are often more efficient than the Jacobian presentation, in that fewer generators or relations are needed. This method has been used to construct infinite cyclic covers of any link [137], and the maximal abelian and maximal free covers of boundary links [63]. The latter construction works equally well for homology boundary links, using "singular Seifert surfaces" [73]. Recently Cooper has shown that the maximal abelian cover of any 1-link can be constructed in a similar way, on using Seifert surfaces which intersect in a controlled manner [34]. (This idea is also implicit in Bailey's thesis [7]).

We shall sketch the construction of the maximal free cover of an homology boundary link. A map $f:G(L) \to F(\mu)$ corresponds to a map $F:X(L) \to \overset{\mu}{\vee} S^1 = K(F(\mu),1)$, which may be assumed transverse to $\{P_1,\ldots,P_\mu\}$, where P_i is a point in the ith copy of S^1 distinct from the wedge point. Let $W_i = F^{-1}(P_i)$ and let $Y = X-M$ where M is an open regular neighbourhood of $W = \cup W_i$. There are two embeddings i_+ and i_- of W in ∂Y and

$X^\omega = Y \times F(\mu)/ < i_+(w_j),\ h.x_j > \sim < i_-(w_j),h >$ for all w_j in W_j, and h in $F(\mu)$.

Here x_j is a generator of $F(\mu)$ corresponding to a loop in X which meets W_j transversally in one point and avoids the other components of W. (This is just the pull back of the corresponding construction of the universal cover of $\overset{\mu}{\vee} S^1$). There is then a Mayer-Vietoris sequence:

$$\ldots \to \mathbb{Z}[F(\mu)] \otimes H_q(W) \xrightarrow{d_q} \mathbb{Z}[F(\mu)] \otimes H_q(Y) \longrightarrow H_q(X^\omega;\mathbb{Z}) \to \ldots$$

where $d_q(\gamma \otimes v_j) = \gamma\, x_j \otimes (i_{j+})_*(v_j) - \gamma \otimes (i_{j-})_*(v_j)$ for γ in $\mathbb{Z}[F(\mu)]$ and v_j in $H_q(W_j;\mathbb{Z})$. If L is a boundary link we may assume that W_j are Seifert surfaces; if L is also a 1-link then since the longitudes bound the W_j, which lift to X^ω, they must be null homologous there, and so lie in $(G_\omega)'$. The Mayer-Vietoris sequence shows that for a boundary n-link $H_{n+1}(X^\omega;\mathbb{Z})$ is a free $\mathbb{Z}[F(\mu)]$-module of rank $\mu - 1$. If L is a boundary 1-link then $H_1(X^\omega;\mathbb{Z}) = G_\omega/G_\omega'$ has a presentation with equal numbers of generators and of relations. (This remains true if L is an homology boundary link whose longitudes lie in G_ω').

A similar construction of the maximal abelian cover of the exterior of a boundary n-link together with an interpretation of the maps d_q in terms of Alexander duality in S^{n+2} (as given by Levine for knots [116]) leads to the conclusion that these maps are monomorphisms if $q \geq 1$ and hence that $p.d.H_q(X;\Lambda) \leq 2$ for all $q \geq 2$. For boundary 1-links $H_2(X;\Lambda)$ is free of rank $\mu - 1$. (Sato has also investigated the homology of the maximal abelian cover of the exterior of a boundary n-link [162 , 163]).

CHAPTER II RIBBON LINKS

In this chapter we introduce the class of links that shall provide most of our examples, and we establish several properties useful in the verification of these examples.

Definition A μ-component n-ribbon map is an immersion (local embedding) $R : \mu D^{n+1} \to S^{n+2}$ with no triple points and such that the components of the singular set are n-discs whose boundary (n-1)-spheres either lie on $\mu S^n = \partial(\mu D^{n+1})$ ("throughcut") or are disjoint from μS^n ("slit"). A μ-component n-link is a ribbon link if there is an n-ribbon map R such that $L = R|\partial(\mu D^{n+1})$.

A ribbon 1-link may be depicted schematically as in Figure 1.

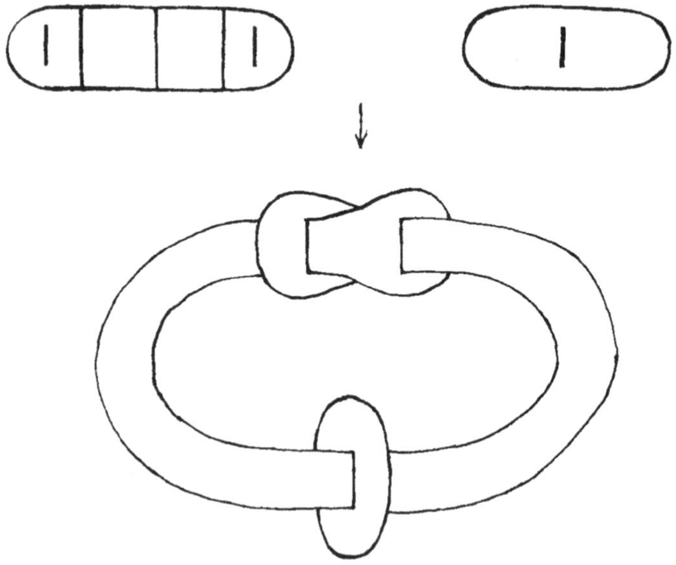

Figure 1

It is easy to see that if L is a ribbon link, the ribbon R may be deformed so that each component of the complement of the throughcuts is bounded by at most two throughcuts. In what follows R will always be so chosen.

It is well known and easy to see that ribbon links are null concordant [49]. The converse remains an open conjecture when $n = 1$, even for knots [50 ; Problem 25], but is false in higher dimensions [81, 206] as will be shown below.

Theorem 1. Let L be a μ-component ribbon n-link. Then L is a sublink of a ν-component ribbon n-link $\hat{L}$ for which surgery on the longitudes gives $\#^{\nu}(S^1 \times S^{n+1})$. In particular $\hat{L}$ is an homology boundary link.

Proof. Let $R : \mu D^{n+1} \to S^{n+2}$ be a ribbon extending L. Let S_i, $1 \leq i \leq \sigma$, be the slits of R and for each slit choose a regular neighbourhood N_i contained in the interior of the corresponding disc and such that $N_i \cap N_j = \phi$ for $i \neq j$. Let $\nu = \mu + \sigma$ and let $\hat{L} = R|(\mu S^n \cup \bigcup_{1 \leq i \leq \sigma} \partial N_i)$. Clearly $\hat{L}$ is a ν-component ribbon n-link with L as a sublink. If $n > 1$ the normal bundle of $\hat{L}$ in S^{n+2} has an essentially unique framing; if $n = 1$ give each component of $\hat{L}$ the 0-framing. Let $W(\hat{L}) = D^{n+3} \cup_T \nu D^{n+1} \times D^2$ where $T : \nu S^n \times D^2 \to S^{n+2}$ is an embedding of a regular neighbourhood of $\hat{L} = T|\nu S^n \times \{0\}$ determined by this framing. Then $\partial W(\hat{L})$ is the result of surgery on S^{n+2} along the longitudes of $\hat{L}$.

Now by adding a pushoff of $\hat{L}|\partial N_i$ to the component of L bounding the (n+1)-disc containing N_i, $\hat{L}$ may be replaced by a ribbon link with one less singularity; moreover if $n = 1$ each component of the new link still has the 0-framing. Continuing thus $\hat{L}$ may be replaced by a ribbon link $\tilde{L}$ for which the only singularities are those corresponding to the components ∂N_i. Clearly these components may be slipped off the ends of the other components of the new ribbon and so $\tilde{L}$ is a trivial ν-component link. Now adding pushoffs of link components to one another (a Kirby move of type 2 [105]) corresponds to sliding $(n+1)$-handles of $W(\hat{L})$ across one another, which leaves unchanged the topological type of $W(\hat{L})$ and hence of $\partial W(\hat{L})$. Thus $\partial W(\hat{L})$ is homeomorphic to $\partial W(\tilde{L}) = \overset{\nu}{\#}(S^1 \times S^{n+1})$ and $G(\hat{L})$ maps onto $\pi_1(\partial W(\hat{L})) \approx F(\nu)$. //

If $n = 1$, the kernel of the map $G(\hat{L}) \to F(\nu)$ is necessarily $G(\hat{L})_\omega$, and is trivial if and only if $\hat{L}$ is trivial, in which case L is also trivial. If $n > 1$ the map $G(\hat{L}) \to F(\nu)$ is an isomorphism, but need not carry any set of ν meridians to a basis. This is so if and only if $\hat{L}$ is a boundary link, in which case $\hat{L}$ (and hence L) is trivial, if $n > 2$, by Gutiérrez' unlinking theorem [61], since $\pi_i(S^{n+2} - \operatorname{im}\hat{L}) \approx \pi_i(S^{n+2} - \operatorname{im}\tilde{L}) = 0$ for $2 \leq i < n$. The above theorem is not the best possible, in that fewer new components may suffice to trivialize L thus. For instance, if L is the square knot, (figure 1 shows that) it is a component of a 2-component homology boundary link with the above property. Recalling that a 1-knot is said to have Property R if surgery on a longitude of the knot does <u>not</u> give $S^1 \times S^2$, and that it has been conjectured that all non trivial 1-knots have Property R [106; Problem 1.16], this example shows that the most direct analogue of Property R for links fails already for a 2-component link. Is there a non trivial boundary 1-link for which 0-framed surgery gives a connected sum of copies of $S^1 \times S^2$?

Kirby and Melvin showed that any knot which does not have Property R is (TOP) null concordant [107], and this suggests the following complement to the above result.

Theorem 2 *If $n \geq 2$ and L is a ν-component n-link such that surgery on the longitudes of L gives $\overset{\nu}{\#}(S^1 \times S^{n+1})$, then L is an homology boundary link and is null concordant. (Hence also any sublink of L is null concordant).*

Proof. That L is an homology boundary link is clear. Let U(L) be the trace of the surgeries on L, so $\partial U(L) = S^{n+2} \amalg \overset{\nu}{\#}(S^1 \times S^{n+1})$. Then $D(L) = U(L) \cup \overset{\nu}{\natural}(D^2 \times S^{n+1})$ is a contractible $(n+3)$-manifold with boundary S^{n+2}, and so is an $(n+3)$-disc. The link L clearly bounds ν disjoint $(n+1)$-discs in D(L). //

If $n = 1$ it can be proven that L bounds ν embedded discs in a contractible 4-manifold W_o, by imitating the first part of the theorem of Kirby and Melvin. Whether the Mazur trick may be used to show that W_o is D may be related to the Andrews-Curtis conjecture [106 ; Problem 5.2]. This comment is due to Rubinstein, who has also recently proven that if 0-framed surgery on the longitudes of the first ρ components of L gives $\overset{\rho}{\#}(S^1 \times S^2)$, for each $\rho \leq \nu$, then L is TOP null concordant [160].

In higher dimensions links which are not homology boundary links but which are sublinks of homology boundary links may be constructed as a consequence of the following theorem.

Theorem 3. A finitely presentable group G is the group of a μ-component sublink of an n-link $L : \nu S^n \to S^{n+2}$ with group free (for some ν and any $n \geq 2$), if and only if G has a presentation of deficiency μ and is normally generated by μ elements. (If $n = 2$ the ambient space may be merely a homotopy 4-sphere.)

Proof. The necessity of the condition is obvious. Suppose that G has a presentation $\{X_i,\ 1 \leq i \leq \nu \mid r_j\ 1 \leq j \leq \nu-\mu\}$ and that s_k, $1 \leq k \leq \mu$, are words in the symbols X_i whose images generate G normally. The fundamental group of $\#^{\nu}(S^1 \times S^{n+1})$ is isomorphic to $F(\nu)$, and the words r_j and s_k may be represented by embeddings $\rho_j : S^1 \to \#^{\nu}(S^1 \times S^{n+1})$ and $\sigma_k : S^1 \to \#^{\nu}(S^1 \times S^{n+1})$ respectively. If surgery is performed on all the ρ_j, $1 \leq j \leq \nu-\mu$, and σ_k, $1 \leq k \leq \mu$, then the resulting manifold is a homotopy $(n+2)$-sphere, and

$$\#^{\nu}(S^1 \times S^{n+1}) - \bigcup^{\nu-\mu} \rho_j(S^1 \times D^{n+1}) - \bigcup^{\mu} \sigma_k(S^1 \times D^{n+1})$$

is the complement of a ν-component n-link in this homotopy sphere with fundamental group $F(\nu)$ [96]. Therefore if surgery is performed only on the ρ_j, $1 \leq j \leq \nu-\mu$, the space

$$((\#^{\nu}(S^1 \times S^{n+1}) - \bigcup^{\nu-\mu} \rho_j(S^1 \times D^{n+1})) \cup \bigcup^{\nu-\mu} (D^2 \times S^n)) - \bigcup^{\mu} \sigma_k(S^1 \times D^{n+1})$$

is the complement of a μ-component sublink with group presented by $\{X_i,\ 1 \leq i \leq \nu \mid r_j,\ 1 \leq i \leq \nu-\mu\}$, that is, with group G. //

If for instance G_{ij} is the group with presentation $\{X_1,X_2,X_3 \mid X_1^{-1}[X_3^i,X_1][X_3^j,X_2]\}$ where $ij \neq 0$, then according to Baumslag [10], G_{ij} is parafree but not free, and so cannot map onto $F(2)$. Thus the link constructed as above from this presentation is not an homology boundary link, although it is a sublink of a 3-component homology boundary link. (Notice that G_{ij} is the normal closure of the images of X_2 and X_3, and the presentation $\{X_1,X_2,X_3 | X_1^{-1}[X_3^i,X_1][X_3^j,X_2],X_2,X_3\}$ of the trivial group is AC-equivalent to the trivial presentation, so this group can be realized by a 2-link in S^4 [120]). We shall show in Chapter V that a ribbon 1-link need not be an homology boundary link, although by Theorem 3 it is a sublink of one. In contrast a sublink of a boundary link is always a boundary link.

An immediate consequence of Theorems 1 and 3 is that if $n > 1$ the group of a μ-component ribbon n-link has a presentation of deficiency μ. Therefore for instance Fox's 2-knot with non principal first Alexander ideal [49] is slice [96] but not ribbon. (This was shown earlier by Hitt [81] and Yanagawa [206]).

Let R be a ribbon map extending the 1-link L. We shall show that although the group of an unsplittable 1-link has no presentation of deficiency greater than 1, in the case of a ribbon 1-link such a ribbon R determines a quotient of the link group which has deficiency μ. Each throughcut T of R determines a conjugacy class $g(T) \subset G(L)$ represented by the image of the oriented boundary of a small disc neighbourhood in R of the corresponding slit. (The standard orientation of D^2 induces an orientation on this neighbourhood via the local homeomorphism R.)

<u>Definition</u> The <u>ribbon group</u> of R is the group

$$H(R) = G(L) / \langle\langle \cup g(T) \mid T \text{ a throughcut of } R \rangle\rangle .$$

(Recall that $\langle\langle S \rangle\rangle$ denotes the smallest normal subgroup of G containing S.)

Lemma 4 The longitudes of L are in $<< \cup g(T) \mid T$ a throughcut of $R >>$.

Proof   Each longitude is represented (up to conjugacy) by a curve on and near the boundary of the corresponding disc, which is clearly homotopic to a product of (conjugates of) loops about the slits in that disc. (See Figure **2**). //

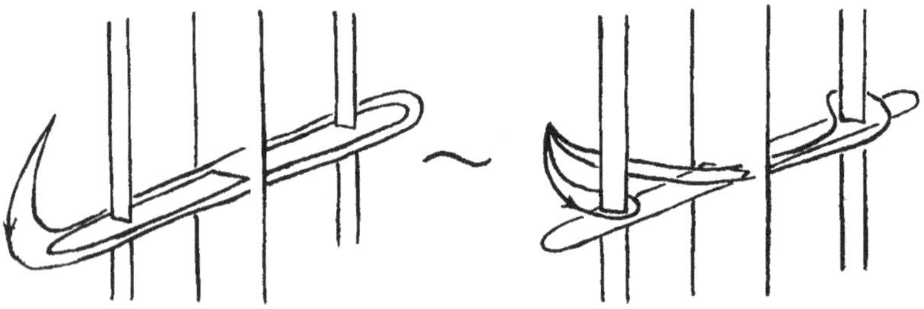

Figure 2

Lemma 5   For all throughcuts T, $g(T) \subset G_\omega$.

Proof   Certainly, for all T, $g(T) \subset G_1$.   Suppose all $g(T) \subset G_n$.   Then their images $\overline{g(T)}$ are central in G/G_{n+1}.   It then follows that $\overline{g(T)} = \overline{g(T')}$ where T' is either throughcut adjacent to T, and hence, moving along the ribbon, that all $\overline{g(T)} = \{1\}$, in other words that all $g(T) \subset G_{n+1}$.   (See Figure 3).   By induction, all $g(T) \subset G_\omega$. //

$$x = sys^{-1} \equiv y \mod S$$
$$\therefore \quad g(T) \equiv g(T') \mod g(S)$$

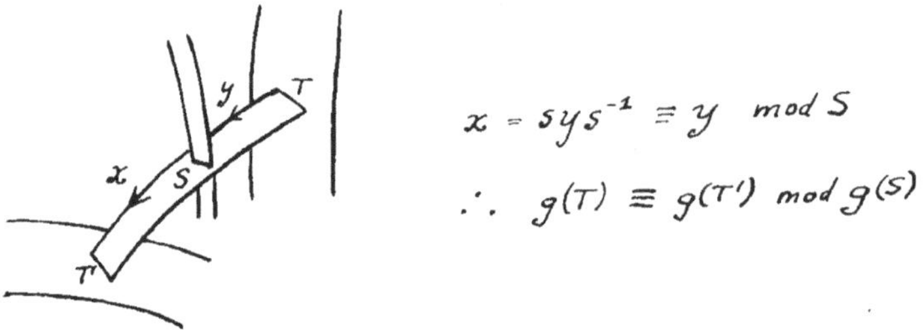

Figure 3

Theorem 6 For any ribbon link L the projection $G \to G/G_\omega$ factors as $G \to G / \ll \text{longitudes} \gg \to H(R) \to G/G_\omega$, and $H(R)$ has a presentation of deficiency μ of the form

$$\{x_{ij},\ 1 \leq j \leq j(i),\ 1 \leq i \leq \mu \mid w_{ij} x_{ij} w_{ij}^{-1} = x_{ij+1},\ 1 \leq j < j(i),\ 1 \leq i \leq \mu\}$$

where there is one generator x_{ij} for each component of the complement of the throughcuts and one word w_{ij} of length one for each throughcut. (Here R is any ribbon map extending L which satisfies the condition imposed after Figure 1).

Proof The factorization of $G \to G/G_\omega$ follows from the lemmas. It may be assumed that in a generic projection of the ribbon there are no triple points. The Wirtinger generators of the link group corresponding to the subarcs of the projection of the link which "lie under" a segment of the ribbon may be deleted, and the two associated relations replaced by one stating that either adjacent generator is conjugate to the other by a loop around the overlying segment. (See Figure 4).

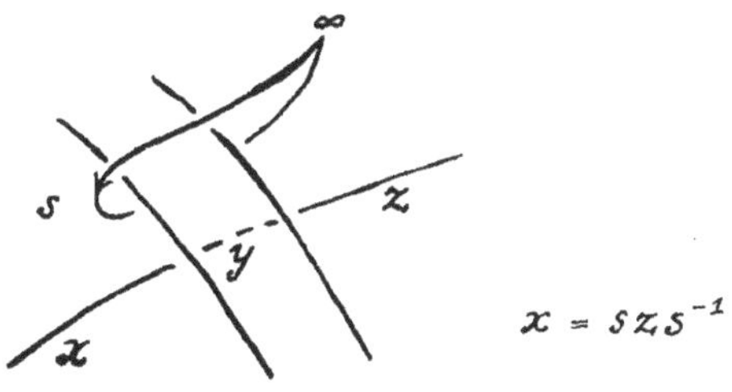

Figure 4

Any loop about a segment of the ribbon is killed in H(R), for the only obstructions to deforming it onto a loop around the throughcut at an end of the segment are elements in the conjugacy classes of the throughcuts between the loop and that end. (See Figure 5).

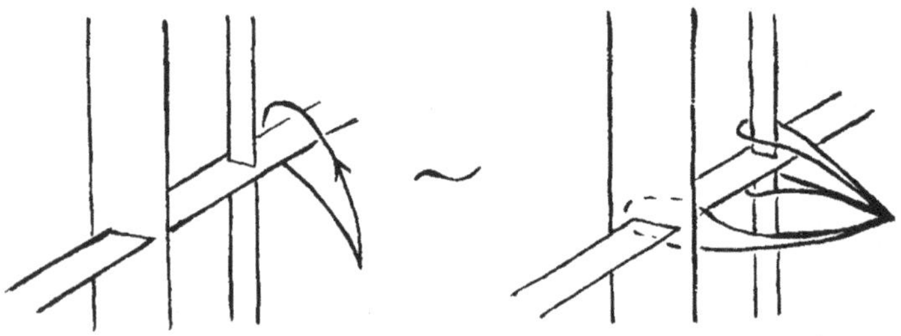

Figure 5

Hence the remaining generators corresponding to subarcs of the boundary of a given component of the complement of the throughcuts coalesce in H(R). Conversely the presentation obtained from the Wirtinger presentation by making such deletions and identifications has the enunciated form, and presents a group in which the image of each g(T) is trivial, for the image of g(T) is trivial if and only if the pair of generators corresponding to arcs meeting the projection of T are identified. Thus the group is exactly H(R). //

Conversely any such presentation can be realized by some ribbon map $R : \mu D^2 \to S^3$. A similar argument shows that a group G is the group of a μ-component ribbon n-link for $n \geq 2$ if and only if G has a Wirtinger presentation of deficiency μ and $G/G' = \mathbb{Z}^\mu$. The generators correspond to meridianal loops transverse to the components of the complements of the throughcuts, and there is one relation for each throughcut. (This was proven for $n = 2$, $\mu = 1$ by Yajima [205]. See Simon [173] and Suzuki [186] for the connection between abstract Wirtinger presentations and homology.) In Chapter VI we show that Baumslag's group $G_{-1,1}$ has such a presentation.

Most of Theorem 6 can be deduced easily from Theorem 1, by arguing as in Theorem 3 to adjoin $\nu - \mu$ relations to $F(\nu) \approx \hat{G}/\hat{G}_\omega$. In general $G/\langle\langle \text{longitudes} \rangle\rangle$, $H(R)$ and G/G_ω are distinct groups, even when $\mu = 1$. (For example, consider the square knot as the boundary of a ribbon link with two throughcuts.) If one ribbon R_1 is obtained from another R_2 by knotting the ribbon or by inserting full twists, then $H(R_1) = H(R_2)$, as such operations do not change the pattern of the singularities. Can H(R) be characterized link- or group-theoretically?

As a partial answer to the above question, H(R) is the group of a μ-component 2-link of which L is a slice, and where the longitudes clearly die. (By pushing the singularities of the ribbon apart in D^4, a null concordance $C : \mu D^2 \to D^4$ of L is obtained; doubling this nullconcordance gives an embedding $2C : \mu S^2 \to S^4$ of which L is a slice, and the group of this 2-link may be computed to be H(R) by Fox's method [48].) Thus in particular the link of Figure 1 is a slice of a 2-component 2-link with group free. Since this 1-link is not a boundary link (see Chapter VI) the 2-link is nontrivial and so this gives a simple example of the phenomenon first observed by Poenaru [149]. (This example was given by Sumners, from a different viewpoint.) Similarly the knot in Figure 6 is a slice of a 2-knot with group $\mathbb{Z}$. Yanagawa [207] has shown that it is in fact a nontrivial slice of a trivial 2-knot.

Figure 6

He showed there also that a ribbon 2-knot with group $\mathbb{Z}$ is trivial.

In [193] Tristram shows that concordance of μ-component 1-links is generated as an equivalence relation by concordances of the form $L \longrightarrow L +_b \partial R$, where $R : \mu D^2 \longrightarrow S^3 - \text{im}L$ is a ribbon map with image disjoint from that of L, and where $+_b$ denotes (iterated) band connected sum.

CHAPTER III DETERMINANTAL INVARIANTS OF MODULES

Throughout this chapter R shall denote an integrally closed noetherian domain. (Although it would suffice for our applications in the next section to assume further that R be factorial, this would not alter the proofs. In fact as our principal technique is to reduce to the case of a discrete valuation ring by localizing at height one prime ideals, most of our results may be extended to the case of an arbitrary Krull domain.)

Let M be a finitely generated R-module. The rank of M over R is the dimension of the vector space $M_o = R_o \otimes_R M$ over R_o, the field of fractions of R. The R-torsion submodule of M is $tM = \{m \text{ in } M \mid r.m = 0$ for some nonzero r in R$\}$, and M is an R-torsion module if tM = 0. The annihilator ideal of M is $\mathrm{Ann}\, M = \{r \text{ in } R \mid r.m = 0 \text{ for all } m \text{ in } M\}$. Let

$$R^p \xrightarrow{Q} R^q \xrightarrow{\phi} M \longrightarrow 0$$

be a finite presentation for M. This presentation has deficiency $q-p$, and is said to give a short free resolution of M if the map Q is injective. For each $k \geq 0$ the kth elementary ideal of M is the ideal $E_k(M)$ generated by the $(q-k)\times(q-k)$ subdeterminants of the matrix representing Q if $k < q$ and by 1 if $k \geq q$. (We use the terminology of Crowell and Fox [43 ; page 101]. This ideal is called the kth determinantal ideal by Bourbaki [13 ; page 573] and the $(k+1)^{st}$ Fitting ideal by Buchsbaum and Eisenbud [19]. That it depends only on M, not on the presentation is well known, and is proven for instance by Crowell and Fox.) Clearly $E_k(M) \subseteq E_{k+1}(M)$ and $E_k(M_S) = E_k(M)_S$ for any multiplicative system S in R.

For each $k \geq 0$ let $\wedge_k M$ be the kth exterior power of M [14 ; page 507] and let $\alpha_k M = \text{Ann } \wedge_k M$. The notation $\alpha_k M$ is due to Auslander and Buchsbaum [5] who showed that if R is local and $\alpha_k M$ is principal for all k then M is a direct sum of cyclic modules, and used this to give criteria for projectivity. Since $\wedge_k (M)_S = (\wedge_k M)_S$ [13; page 78] it follows that $\alpha_k (M_S) = (\alpha_k M)_S$,) while clearly $\alpha_k M \subseteq \alpha_{k+1} M$. In the next result, relating these ideals to the elementary ideals, we shall invoke Cramer's rule in the following form. Let A be an $a \times a$ R-matrix and let $d \neq 0$ divide each of the $(a-1)\times(a-1)$ subdeterminants of A. If u is an $a \times 1$ column matrix (respectively, a $1 \times a$ row matrix) then $((\det A)/d)\, u$ is an R-linear combination of the columns (respectively, rows) of A. See [14; page 535] .

<u>Theorem 1</u> <u>Let M be a finitely presentable R-module. Then</u>

(i) $E_0(M) \subseteq \text{Ann } M = \alpha_1(M)$;

(ii) $\sqrt{E_k}(M) = \sqrt{\alpha_{k+1}}(M)$ <u>for each</u> $k \geq 0$.

<u>Proof</u> (i) We may clearly suppose $E_0(M) \neq 0$. Let D be a $q \times q$ submatrix of a presentation matrix Q for M (as above), with $\delta = \det D$ non-zero. Then by Cramer's rule $\delta \cdot R^q \subseteq D(R^q) \subseteq Q(R^p)$ and so $\delta \cdot \phi(u) = \phi(\delta \cdot u) \in \text{im } \phi \cdot Q = 0$ for all $u \in R^q$. Hence $\delta \in \text{Ann } M$ and so every generator of $E_0(M)$ is in Ann M.

(ii) Let $\wp$ be a prime ideal of R. We must show that $E_k(M) \subseteq \wp$ if and only if $\alpha_{k+1}(M) \subseteq \wp$. We may localize with respect to $R - \wp$ and thus assume that $\wp$ is the unique maximal ideal of R. Let q be the dimension of the vector space $M/\wp M$ over the field $R/\wp$. Then

$\alpha_k(M/\wp M) = 0$ if $k \leq q$ and $\alpha_{q+1}(M/\wp M) = R/\wp$, so $\alpha_k(M) \subseteq \wp$ if and only if $k \leq q$. By Nakayama's lemma [4; page 21], M has a presentation with q generators. Since $M/\wp M$ has dimension q, all the entries of the presentation matrix are in $\wp$, and hence $E_k(M) \subseteq \wp$ if and only if $k < q$, that is, if and only if $\alpha_{k+1}(M) \subseteq \wp$. In other words $\sqrt{E_k}(M) = \cap\{\wp \text{ prime} \mid E_k(M) \subseteq \wp\} = \cap\{\wp \text{ prime} \mid \alpha_{k+1}(M) \subseteq \wp\} = \sqrt{\alpha_{k+1}}(M)$. //

These results are well known. (See for instance [13 ; page 573]). In [19], Buchsbaum and Eisenbud show also that for each $k \geq 0$ $E_k(M) \subseteq \alpha_{k+1}(M) \subseteq (E_k(M) : E_{k+1}(M))$ and give sufficient conditions for this inclusion to be an equality. (Their methods apply to modules over any commutative noetherian ring.)

<u>Definition</u> [13; page 476] The <u>divisorial hull</u> $\tilde{I}$ of an ideal I of R is the intersection of the principal ideals of R which contain I.

It is clear that if S is a multiplicative system in R then $(I_S)^{\sim} = (\tilde{I})_S$ as ideals of the localization R_S, while if R is factorial and $I \neq 0$ the ideal $\tilde{I}$ is a principal ideal, generated by the highest common factor of the elements of I.

<u>Lemma 2</u> <u>The divisorial hull of an ideal</u> I <u>is</u> $\cap I_p$, <u>the intersection of all of its localizations at height one prime ideals</u> p <u>of</u> R.

<u>Proof</u> Since R is an integrally closed noetherian domain, it is a Krull domain, so $R = \cap R_p$ is the intersection of all of its localizations at height one primes, and these are each discrete valuation rings [13 ; pp.480-485]. Therefore if I is an ideal, $(I_p)^{\sim} = I_p$ so $\tilde{I} \subseteq \cap(\tilde{I})_p = \cap(I_p)^{\sim} = \cap I_p$. On the other hand if $I \subseteq (a)$ then $\cap I_p \subseteq \cap(a)_p = (a).\cap R_p = (a)$, so $\cap I_p \subseteq \tilde{I}$. //

If R is factorial and M is a finitely generated R-module, let $\Delta_k(M)$ be any generator of the principal ideal $E_k(M)^{\sim}$, for each $k \geq 0$.

<u>Lemma 3</u> <u>If</u> R <u>is a discrete valuation ring and</u> M <u>is a finitely generated</u> R-<u>module of rank</u> r, <u>then</u> $\alpha_k M = 0$ <u>if</u> $k \leq r$ <u>and</u> $\alpha_{r+j}M = \alpha_j tM = (\Delta_{r+j-1}(M)/\Delta_{r+j}(M))$ <u>for each</u> $j \geq 1$.

<u>Proof</u> Let p be the maximal ideal of R. By the structure theorem for finitely generated modules over principal ideal domains, $M \approx R^r \oplus tM \approx R^r \oplus (\bigoplus_{1 \leq i \leq n} (R/p^{e(i)}))$ where $0 < e(i) \leq e(i+1)$ for $1 \leq i \leq n$.

Therefore $E_k(M) = 0$ if $k < r$ and $E_{r+j}(M) = E_j(tM) = p^{s_j}$ where $s_j = \sum_{1 \leq i \leq n-j} e(i)$ for each $j \geq 0$. Moreover

$$\Lambda_k M \approx \bigoplus_{0 \leq j \leq k} (\Lambda_j(R^r) \otimes \Lambda_{k-j} tM) = \bigoplus_{0 \leq j \leq k} (\Lambda_{k-j} tM)^{\binom{r}{j}}.$$

[14;pages 515-518] and hence $\alpha_k M = 0$ if $k \leq r$ and $\alpha_{r+j}M = \alpha_j tM$ for each $j \geq 1$. Therefore we may assume $r = 0$. Then

$$\Lambda_k M \approx \bigoplus_{1 \leq i(1) < .. < i(k) \leq n} ((R/p^{e(i(1))}) \otimes . . \otimes (R/p^{e(i(k))}))$$

$$\approx \bigoplus_{1 \leq i(1) < .. < i(k) \leq n} (R/p^{e(i(1))})$$

$$\approx \bigoplus_{1 \leq i \leq n} (R/p^{e(i)})^{f(i)}$$

where $f(i) = \mathrm{Card}\{(i(2),\ldots i(k)) \text{ in } \mathbb{Z}^{k-1} \mid i < i(2) < .. < i(k) \leq n\}$. Clearly $f(i) = 0$ if $i > n-k+1$ and $f(n-k+1) = 1$. Therefore

$$\alpha_k M = \mathrm{Ann}(\Lambda_k M) = p^{e(n-k+1)} = (\Delta_{k-1}(M)/\Delta_k(M)). //$$

If we note that the above ideal $(\Delta_{r+k-1}(M)/\Delta_{r+k}(M))$ is the ideal quotient $(E_{r+k-1}(M) : E_{r+k}(M)) = \{S \text{ in } R | S.E_{r+k}(M) \subseteq E_{r+k-1}(M)\}$ when R is a discrete valuation ring, then we may extend this result as follows.

__Theorem 4__ __If__ M __is a finitely generated__ R-__module, then__ $\alpha_k M = 0$ __for each__ $k \leq r = \text{rank } M$ __and__ $(\alpha_{r+j}M)^{\sim} = (E_{r+j-1}(M) : E_{r+j}(M))^{\sim}$ __for each__ $j \geq 1$.

__Proof__ By Lemma 2 it is enough to prove that the localizations of the ideals at height one primes are the same. This follows from Lemma 3 on observing that every step (forming exterior powers, annihilators etc.) is compatible with localization. //

__Corollary__ __If the coefficient ring is factorial, then__ $(\alpha_{r+j}M)^{\sim} = (\Delta_{r+j-1}(M)/\Delta_{r+j}(M))$ __for each__ $j \geq 1$. __Hence__ $(\Delta_{r+j}(M)) = (\Delta_j(tM))$ __for each__ $j \geq 0$.

__Proof__ By Lemma 3 $(\alpha_{r+j}M)^{\sim} = (\alpha_j tM)^{\sim}$ for each $j \geq 1$. Now apply the theorem. //

__Remarks__ 1. The second part of this Corollary was proven by Blanchfield [11 ; Lemma 4.10].

2. If M is a torsion module (over a factorial domain) which has a square presentation matrix then it follows easily from Cramer's rule that $(\Delta_0(M)/\Delta_1(M)) \subseteq \text{Ann } M$. The Corollary then implies that $\text{Ann } M = (\Delta_0(M)/\Delta_1(M))$. (Buchsbaum and Eisenbud show that if R is any noetherian ring and M is an R-module with a square presentation matrix whose determinant is not a zero divisor, then $\text{Ann } M = (E_0(M):E_1(M))$ [19]).

Lemma 5 Let $0 \to A \to B \to C \to 0$ be an exact sequence of R-modules such that A is an R-torsion module and rank B = rank C = r. Then

$E_r(B)^{\sim} = E_0(A)^{\sim}E_r(C)^{\sim}$.

Proof If R is a principal ideal domain, this is an immediate consequence of the structure theorem for R-modules. The general case then follows on localization at height one primes.//

This lemma was first proven by Levine [117] for R factorial and A, B and C all R-torsion modules. If C has a short free resolution we can do slightly better.

Lemma 6 Let $0 \to K \to M \to Q \to 0$ be a short exact sequence of R-modules. Then $E_i(M) \supseteq E_j(K)E_{i-j}(Q)$ for all $i \geq j$. If Q has a short free resolution and is of rank r, then $E_r(M) = E_0(K).E_r(Q)$.

Proof Given presentation matrices P(K) and P(Q) for K and Q respectively it is easy to see that M has a presentation matrix of the form $\begin{pmatrix} P(K) & 0 \\ * & P(Q) \end{pmatrix}$. (Here we assume the columns correspond to the generators and the rows to the relations.) Hence $E_i(M) \supseteq E_j(K)E_{i-j}(Q)$ for all $i \geq j$. If Q has a short free resolution we may assume that P(Q) is a $q \times (q+r)$ matrix, where r = rank Q. It is then easy to see that the only nonzero elements of $E_i(M)$ are those obtained by deleting r columns form P(Q) and taking the product of the resulting element of $E_r(Q)$ with a subdeterminant of P(K) of column index 0. //

Definition [13; page 523] An R-module M is pseudozero if $M_p = 0$ for every height one prime ideal p of R.

Lemma 7 i) *M is pseudozero if and only if* $(\text{Ann } M)^{\sim} = R$.

ii) *If* N *is a pseudozero submodule of* M, *then* $(\alpha_k M)^{\sim} = (\alpha_k (M/N))^{\sim}$ *and* $E_k(M)^{\sim} = (E_k(M/N))^{\sim}$ *for each* $k \geq 0$.

Proof Since $(\alpha_k M)_p = \alpha_k(M_p)$ and $E_k(M)_p = E_k(M_p)$ for each $k \geq 0$, the first assertion follows from Lemma 2 on considering $\alpha_1 M$ and the second follows from the equations $(\alpha_k M)_p = \alpha_k(M_p) = \alpha_k((M/N)_p) = (\alpha_k(M/N))_p$ and $E_k(M)_p = E_k(M_p) = E_k((M/N)_p) = E_k(M/N)_p$. //

Lemma 8 *Let* A *be a* b×c *R-matrix of rank* d, *and suppose there is a* d×d *submatrix* D *such that* det D *divides every* d×d *subdeterminant of* A. *Then there are invertible square matrices* B *and* C *such that* $BAC = \begin{pmatrix} D & 0 \\ 0 & 0 \end{pmatrix}$

Proof After permuting the rows and columns if necessary, we may assume that D is in the top left hand corner of A. We may then apply Cramer's rule to annihilate the partial rows and columns below and to the right of D. The bottom right hand corner block of the resulting matrix must be null as rank BAC = rank D. //

Recall that a nonzero ideal I of a noetherian domain R is called *invertible* if it is projective as an R-module, and that a finitely generated module M is projective if and only if the localization M_p is a free R_p-module for each prime ideal p of R [13;page 109].

Theorem 9 *Let* M *be a finitely generated* R-*module of rank* r. *Then* $E_r(M)$ *is invertible if and only if* P = M/tM *is projective and* tM *has projective dimension at most one. In this case* $M \approx P \oplus tM$ *and* $E_{r+j}(M) = E_j(tM)$ *for each* $j \geq 0$.

<u>Proof</u> Suppose that $E_r(M)$ is invertible, and that M has a presentation

$$R^p \xrightarrow{Q} R^q \xrightarrow{\phi} M \longrightarrow 0.$$

Let p be a prime ideal of R. On localizing at p the hypothesis of lemma 8 is satisfied by the matrix Q, for any generating set of a principal ideal in a local domain must contain a generator, by Nakayama's lemma . Thus $M_p \approx R_p{}^r \oplus tM_p$, where tM_p has a short free resolution

$$0 \longrightarrow R_p^{q-r} \xrightarrow{D} R_p^{q-r} \longrightarrow tM_p \longrightarrow 0,$$

and $P_p = M_p/tM_p$ is free. Therefore P_p is projective, so the projection of M onto P = M/tM splits, and if $\psi : R^c \to tM$ is any epimorphism, then ker ψ is locally free (e.g. by Schanuel's lemma [158; page 92]) and hence projective, so p.d.tM $\leq$ 1. Conversely, if P is projective and p.d.tM $\leq$ 1, then P_p is free and tM_p has a short free R_p-resolution, given by a square matrix since tM_p is a torsion module. Hence $E_r(M)_p = E_r(M_p) = E_r(P_p \oplus tM_p) = E_o(tM_p)$ is principal and so $E_r(M)$ is invertible. The last assertion is clear. //

<u>Corollary</u> [18; Lemma 1] M <u>is projective if and only if</u> $E_r(M) = R$.

<u>Proof</u> This follows from the fact that tM = 0 if and only if $E_o(tM) = R$. //

<u>Theorem 10</u> <u>If</u> M <u>is a finitely generated</u> R-<u>module of projective dimension</u> 1, <u>then</u> M <u>has no nontrivial pseudozero submodules</u>.

<u>Proof</u> Let N be a pseudozero submodule of M and let p be a prime ideal of R. Then N_p is a pseudozero submodule of M_p. Since N = 0 if and only if $N_p = 0$ for all p, and since localization is exact, we may therefore assume that R is local, and hence that M has a short free resolution

$$0 \longrightarrow R^p \xrightarrow{Q} R^q \xrightarrow{\phi} M \longrightarrow 0.$$

We shall proceed by induction on $r = q-p$, the rank of M.

If $r = 0$ then $E_o(M)$ is principal and so $E_o(M)^\sim = E_o(M)$. Since N is pseudozero, $E_o(M/N)^\sim = E_o(M)^\sim$ by Lemma 7 and so the inclusions $E_o(M) \subseteq E_o(M/N) \subseteq E_o(M/N)^\sim$ are all equalities. Since M/N is a quotient of M it has a presentation matrix of the form $\begin{pmatrix} Q \\ Q_1 \end{pmatrix}$ which by the argument of Lemma 8 may be changed to $\begin{pmatrix} Q \\ 0 \end{pmatrix}$ by row operations. Hence the projection of M onto M/N is an isomorphism, so $N = 0$.

Suppose $r \geq 1$. Then we may assume that $\phi(e_q)$ generates a free submodule of M. (Here $\{e_1, \ldots e_q\}$ denotes the standard basis of R^q.) Let $M' = M/R.\phi(e_q)$ and let $f: M \to M'$ be the natural projection. Then M' has a short free resolution and its rank is $r-1$, so by the inductive hypothesis M' has no nontrivial pseudozero submodule. But f maps N monomorphically to M', as $R.\phi(e_q)$ is torsion free. Therefore $N = 0$ and the theorem is proven. //

<u>Remark</u> If R is factorial and M is a torsion module then $E_o(M)$ principal implies that M has no nontrivial pseudozero submodule (by Theorems 9 and 10), and since $(\mathrm{Ann}\ M)^\sim.M$ is certainly pseudozero this implies that Ann M is principal. However in general these implications are strict. For instance if $R = \mathbb{Z}[x,y]$ then $R/(x) \oplus R/(x,y)$ has annihilator ideal principal while its second summand is pseudozero, and any maximal ideal of $\mathbb{Z}[x]$ considered as an R-module in the obvious way has no nontrivial pseudozero submodule but its 0th elementary ideal is not principal.

Going in the other direction we may characterize pseudozero modules homologically. For each $k \geq 0$, let e^kM denote $\mathrm{Ext}_R^k(M,R)$. The module $e^oM = \mathrm{Hom}_R(M,R)$ is torsion free, and there is a natural "evaluation" homomorphism $ev_M : M \to e^o e^o M$, with kernel tM. Therefore M is a torsion module if and only if $e^o M = 0$. If M is a torsion module then $e^1 M = Hom_R(M, R_o/R)$ (as follows on applying $Ext_R^*(M,-)$ to exact sequence $0 \to R \to R_o \to R_o/R \to 0$) and there is an "evaluation" homomorphism $W_M : M \to e^1 e^1 M$.

<u>Theorem 11</u> <u>Let M be a finitely generated R-module. Then $Hom_R(M,R_o/R)$ is a torsion module with no nontrivial pseudozero submodule, and ker W_{tM} is the maximal pseudozero submodule of M. Hence M is pseudozero if and only if $e^o M = e^1 M = 0$.</u>

<u>Proof</u> Let $m_1, .. m_g$ generate M and let $f : M \to R_o/R$. If $f(m_i) = r_i/s_i$ mod R (with $s_i \neq 0$) then $s.f = 0$ with $s = \prod_{i=1}^{g} s_i \neq 0$, and so $Hom_R(M,R_o/R)$ is a torsion module. If $f_p = 0$ for all height one prime ideals p then for each m in M f(m) is in the intersection of all the localizations R_p and hence is in R, since R is a Krull domain. Therefore $Hom_R(M,R_o/R)$ has no nontrivial pseudozero submodule. It is an immediate consequence of the structure theorem for finitely generated modules over a principal ideal domain that $W_{tM,p} : tM_p \to e^1 e^1 tM_p$ is an isomorphism. Hence $\ker W_{tM}$ is pseudozero and the remaining assertions follow readily. //

<u>Remarks</u> 1 Theorem 11 implies that if M is a finitely generated torsion module supporting a nondegenerate bilinear pairing $b : M \times M \to R_o/R$ then M has no nontrivial pseudozero submodule. This was proven by Blanchfield [11].

2 It is proven in [113] that if R is a regular local ring of dimension d and M is a finitely generated R-module then M has no nontrivial submodule of finite length if and only if $e^d M = 0$. By localizing at

maximal ideals, it follows that this remains true for modules over a noetherian domain all of whose localizations at maximal ideals are regular of dimension d.

We conclude this chapter by considering another determinantal invariant. Let M be an R-module with a finite presentation

$$R^p \xrightarrow{Q} R^q \xrightarrow{\phi} M \longrightarrow 0$$

and suppose that rank M = r. Then the matrix Q is of rank q-r and the $(q-r)^{th}$ compound $Q^{(q-r)}$ is of rank 1. Steinitz [179] and Fox and Smythe [51] showed that the ideal class of the ideal generated by the elements of any one column (respectively, row) of $Q^{(q-r)}$ depends only on the module M. (Two nonzero ideals I and J belong to the same <u>ideal class</u> if there are nonzero elements a and b in R such that aI = bJ.) Let $\gamma(M)$ and $\rho(M)$ denote the column and row ideal classes of M, respectively. It is easy to see that two nonzero ideals are in the same ideal class if and only if they are isomorphic as R-modules, and that every finitely generated torsion free R-module of rank 1 is isomorphic to an ideal. With this in mind, the Steinitz-Fox-Smythe row invariant may be characterized as follows.

<u>Theorem 12</u> <u>The row ideal class</u> $\rho(M)$ <u>is the isomorphism class of the rank</u> 1 <u>torsion free module</u> $(\Lambda_r M)/t(\Lambda_r M)$.

<u>Proof</u> Let U be a $(q-r)\times q$ submatrix of maximal rank q-r of the presentation matrix Q. Define

$$\psi : (R^q)^r \to R \text{ by } \psi(V_1, \ldots V_1, \ldots V_r) = \det\left[V_1^{tr}, \ldots V_r^{tr}, \ldots V_r^{tr} U^{tr}\right]$$

where the vectors $V_1,\ldots V_r$ in R^q are used as the first r columns of a $q \times q$ matrix. The map ψ is clearly alternating (for if two of the arguments V_i are interchanged the sign of the determinant changes) and $\psi(V_1,\ldots V_r) = 0$ if any of the arguments V_i are in the image of Q, so ψ factors through $\Lambda_r M$. Since R is torsion free there results a map $\bar{\psi}: (\Lambda_r M)/t(\Lambda_r M) \to R$ which clearly has image the ideal generated by the $(q-r) \times (q-r)$ minors of U, which are just the elements of one row of $Q^{(q-r)}$. Since both domain and image of $\bar{\psi}$ are rank 1 torsion free modules, $\bar{\psi}$ gives an isomorphism of $(\Lambda_r M)/t(\Lambda_r M)$ with this ideal. //

The projection of M onto M/tM induces an epimorphism $\Lambda_r M \to \Lambda_r(M/tM)$ and hence $\rho(M) = \rho(M/tM)$. If N is another finitely generated R-module of rank S, then $\rho(M \oplus N)$ is the ideal class of $(\Lambda_{r+s}(M \oplus N))/t(\Lambda_{r+s}(M \oplus N)) = ((\Lambda_r M)/t(\Lambda_r M)) \otimes ((\Lambda_s N)/t(\Lambda_s N))$, and so $\rho(M \oplus N) = \rho(M).\rho(N)$. Thus ρ is a homomorphism from the semigroup of finitely generated R-modules (with respect to direct sum) to the semigroup of ideal classes (with respect to product of ideals). Moreover if F is free $\rho(F) = R$, so $\rho(M)$ depends only on the stable isomorphism class of M ($\rho(M \oplus F) = \rho(M)$) and if P is projective $\rho(P)$ is the class of an invertible ideal.

Let $M^* = \mathrm{Hom}_R(M,R) = \mathrm{Hom}_R(M/tM,R)$. Then M^* is finitely generated, torsion free and rank M^* = rank M = r. There is a natural map $\delta: \Lambda_r(M^*) \to (\Lambda_r M)^*$ given by $\delta(f_1\wedge..\wedge f_r)(m_1\wedge..\wedge m_r) = \det[f_i(m_j)]$, which is an isomorphism if R is a field. The definition of δ is compatible with localization and with passage to a quotient with respect to an ideal. Therefore by Nakayama's lemma δ_p is an epimorphism for each prime ideal p of R. Hence δ is an epimorphism and therefore $\rho(M^*) = \rho(M)^*$.

The column ideal class of $M = \text{coker } Q$ is the row ideal class of $\text{coker } Q^{tr}$, the cokernel of the transpose of Q. The relationship between M and $\text{coker } Q^{tr}$ is rather obscure, but there are a pair of exact sequences

$$(1) \qquad 0 \to M^* \to R^q \xrightarrow{Q^{tr}} R^p \to \text{coker } Q^{tr} \to 0$$

and similarly

$$(2) \qquad 0 \to (\text{coker } Q^{tr})^* \to R^p \xrightarrow{Q} R^q \to M \to 0.$$

If Q_2 is another presentation matrix for M, then it may be related to Q by "Tietze moves" [43] and on examining the effect of such moves on the presentations it may be seen that $\text{coker } Q^{tr}$ and $\text{coker } Q_2^{tr}$ are stably isomorphic. If Q is a monomorphism, so that M has a short free resolution, then $\text{coker } Q^{tr} = e^1 M$ and $\gamma(M) = R$. More generally if $\text{p.d.}M \leq 1$ then $\gamma(M)$ is invertible.

Suppose now that M is projective and let $K = \ker Q = (\text{coker } Q^{tr})^*$. Then the sequence (2) above splits, and hence $\text{coker } Q^{tr} = K^*$ is also projective and the sequence (1) is split exact. Therefore $\gamma(M) = \rho(K^*) = \rho(M^*)$ (since (1) splits) $= \rho(M)^*$. Similarly $\gamma(M^*) = \rho(M)$, since projective modules are reflexive ($M \cong M^{**}$), or in particular because invertible ideals are reflexive. A simple argument due to Steinitz [179] shows that for any finitely generated module M the product $\rho(M).\gamma(M)$ is the class of $E_r(M)$, the ideal generated by <u>all</u> the $(q-r) \times (q-r)$ minors of Q.

It is not generally true that $\gamma(M) = \rho(M)^*$ for every finitely generated module. If $R = Q[x,y,z]$ and M is the ideal (x,y,z) then $\gamma(M) = \rho(M) =$ the class of M, which is not principal, but $M^* = R$ and $\rho(M)^* = R$. It would be interesting to have a simple general characterization of the column invariant of a module and to find other invariants of stable isomorphism classes of (not necessarily projective) modules.

CHAPTER IV THE CROWELL EXACT SEQUENCE

Let L be a μ-component 1-link with exterior X and group G, and let * be a basepoint for X. The long exact sequence of equivariant homology for the covering of the pair (X,*) determined by the maximal abelian cover $p : X' \to X$ gives rise to a 4-term exact sequence of Λ-modules

$$0 \to H_1(X;\Lambda) \to H_1(X,*;\Lambda) \to H_0(*;\Lambda) \to H_0(X;\Lambda) \to 0.$$

As Crowell showed that knowledge of this exact sequence was equivalent to knowledge of the second commutator quotient G/G'' (so that the study of G/G'' could be linearized) we shall call it the Crowell exact sequence. In this chapter we shall establish some results of Crowell, Strauss and Traldi relating the elementary ideals of the members of this sequence, and we shall sketch some results of Massey on the I-adic completion of the sequence. We shall first give new proofs of results of Crowell on annihilators and $\mathbb{Z}$-torsion in knot modules, which extend to the many component case, and we shall consider when the Λ-module G'/G'' admits a square presentation matrix.

Definition The Alexander module of L is the Λ_μ-module $A(L) = H_1(X,*;\Lambda)$. The kth Alexander ideal of L is $E_k(L) = E_k(A(L))$ and the kth Alexander polynomial of L is $\Delta_k(L) = \Delta_k(A(L))$. (The Alexander polynomials are only defined up to multiplication by units.)

The right hand map of the Crowell exact sequence may be identified with the augmentation homomorphism $\varepsilon : \Lambda \to \mathbb{Z}$, while the left hand member is just $H_1(X'; \mathbb{Z})$ considered as a Λ-module via the covering transformations. By the Hurewicz theorem $H_1(X'; \mathbb{Z})$ is isomorphic to G'/G''

as an abelian group. There is an alternative description of the Crowell exact sequence in terms of the link group which we shall give as we shall use it in the next chapter.

Let G be a finitely generated group and let $\varepsilon_G : \mathbb{Z}[G] \to \mathbb{Z}$ be the augmentation homomorphism of the group ring of G. Let $\mathfrak{g} = \ker \varepsilon_G$ be the augmentation ideal, generated by $\{g - 1 \mid g \text{ in } G\}$, and let $\mathfrak{g}_2$ be the ideal generated by $\{w - 1 \mid w \text{ in } G'\}$. The map sending gG' to $(g - 1) + \mathfrak{g}^2$ gives an isomorphism of G/G' with $\mathfrak{g}/\mathfrak{g}^2$. The Alexander module of G, $A(G) = \mathfrak{g}/\mathfrak{g}_2\mathfrak{g}$, is in a natural way a $\mathbb{Z}[G/G']$-module, which is finitely generated and hence finitely presentable. The conjugation action of G/G' on G'/G", given by $gG'.aG'' = g a g^{-1}.G''$ for g in G and a in G', makes G'/G" into a $\mathbb{Z}[G/G']$-module, and there is a natural monomorphism $\delta : G'/G'' \to A(G)$ sending aG" to $(a - 1) + \mathfrak{g}_2\mathfrak{g}$, which has image $\mathfrak{g}_2/\mathfrak{g}_2\mathfrak{g}$. The 4-term exact sequence

$$0 \longrightarrow G'/G'' \xrightarrow{\delta} A(G) \xrightarrow{\phi} \mathbb{Z}[G/G'] \xrightarrow{\varepsilon_{G/G'}} \mathbb{Z} \longrightarrow 0$$

(where ϕ sends $(g - 1) + \mathfrak{g}_2\mathfrak{g}$ to $gG' - 1$) is called the <u>Crowell exact sequence</u> for G. Crowell showed that if $f : H \to K$ induces an isomorphism on abelianization, then the induced map on short exact sequences of groups

$$1 \longrightarrow H'/H'' \longrightarrow H/H'' \longrightarrow H/H' \longrightarrow 1$$

is an isomorphism if and only if the induced map on the Crowell sequences of $\mathbb{Z}[H/H']$-modules is an isomorphism [37]. Several other interpretations of A(G) and of the Crowell sequence are given by Crowell [41], Gamst [53] and Smythe [175].

Thus the Crowell sequence may be written as

$$0 \longrightarrow G'/G'' \longrightarrow A(L) \longrightarrow \Lambda \xrightarrow{\varepsilon} \mathbb{Z} \longrightarrow 0 \qquad (1)$$

and is equivalent to the short exact sequence

$$0 \longrightarrow G'/G'' \longrightarrow A(L) \longrightarrow I_\mu \longrightarrow 0 \qquad (1)'$$

where $I_\mu = \ker \varepsilon = (t_1 - 1, \ldots, t_\mu - 1)$ is the augmentation ideal of Λ_μ. A presentation matrix for $A(L) = A(G(L))$ may be obtained from a presentation for $G(L)$ via the free differential calculus [43], but we shall be content with the information obtainable through homological algebra.

Rank, Projective Dimension and $\mathbb{Z}$-torsion

Since X is a compact bounded 3-manifold with Euler characteristic 0, it is homotopy equivalent to a finite 2-dimensional cell complex with 1 0-cell, $n+1$ 1-cells and n 2-cells, so the equivariant chain complex of X' is chain homotopy equivalent to a finite free complex D_* with $D_o = \Lambda$, $D_1 = \Lambda^{n+1}$, $D_2 = \Lambda^n$ and $D_q = 0$ for $q > 2$. The relative complex of the pair $(X', p^{-1}(*))$ is then chain homotopy equivalent to the complex

$$\ldots \longrightarrow 0 \longrightarrow D_2 \longrightarrow D_1 \longrightarrow 0 \ .$$

Since $H_q(X,*;\Lambda) = H_q(X;\Lambda)$ for all $q \geq 2$, there is an exact sequence

$$0 \longrightarrow H_2(X;\Lambda) \longrightarrow \Lambda^n \xrightarrow{d} \Lambda^{n+1} \longrightarrow A(L) \longrightarrow 0 \ . \qquad (2)$$

<u>Definition</u> The <u>Alexander nullity</u> of L, $\alpha(L)$, is the rank of $A(L)$ as a Λ-module.

It is immediate that $\alpha(L) = \min\{k \mid E_k(L) \neq 0\}$ and is ≥ 1. The complex obtained by tensoring the cellular chain complex of the pair $(X', p^{-1}(*))$ over Λ with $\mathbb{Z}$ is just the cellular chain complex of $(X,*)$, so $A(L) \otimes \mathbb{Z} \approx H_1(X,*;\mathbb{Z}) = \mathbb{Z}^\mu$. Therefore $\varepsilon(E_\mu(L)) = \mathbb{Z}$ and so

$\alpha(L) \leqslant \mu$. The Crowell sequence implies that rank $G'/G'' = \alpha(L) - 1$ and $tG'/G'' = tA(L)$, while the exact sequence (2) implies that $H_2(X;\Lambda)$ is torsion free of rank $\alpha(L) - 1$.

Lemma 1 (Cochran [29]) If $\alpha(L) = 2$, then $H_2(X;\Lambda) \approx \Lambda$.

Proof Let u and v belong to $H_2(X) \subseteq \Lambda^n$. Then since rank $H_2(X;\Lambda) = 1$, there are α and β in Λ such that $\alpha u = \beta v$. We may assume that α and β have no common factor. Since Λ is factorial $v = \alpha w$ for some w in Λ^n, which must actually be in $H_2(X;\Lambda)$ by the exactness of (2) and the fact that Λ^{n+1} is torsion free. Therefore every 2-generator submodule of the finitely generated rank 1 Λ-module $H_2(X;\Lambda)$ is cyclic. The lemma follows easily. //

Cochran's result extended to embeddings of arbitrary finite graphs and was published in [30]. In general $H_2(X;\Lambda)$ is free if and only if the projective dimension of A(L) is at most 2. For if p.d. $A(L) \leqslant 2$, Schanuel's lemma applied to the exact sequence (2) implies that $H_2(X;\Lambda)$ is projective, and Suslin has shown that every projective Λ-module is free [185]. The argument in the other direction is obvious. For a boundary link a Mayer-Vietoris argument shows that $H_2(X;\Lambda)$ is free of rank $\mu - 1$, but the 3-component homology boundary link of Figure V.1 has $A(L) \approx \Lambda_3^3 \oplus (\Lambda_3/(t_1 - 2,\ t_2 + 1,\ t_3 - 1))$ and so for this link $H_2(X;\Lambda)$ is not free. If L is unsplittable X is an Eilenberg-MacLane space K(G,1) for the group G (by the Sphere Theorem) and then $H_q(G';\mathbb{Z}) = H_q(X;\Lambda)$. Thus in particular the commutator subgroup of a classical knot group has trivial integral homology in degree greater than 1.

The Crowell exact sequence for the free group $F(\mu)$ is

$$0 \longrightarrow F(\mu)'/F(\mu)'' \longrightarrow \Lambda^{\mu} \longrightarrow \Lambda \longrightarrow \mathbb{Z} \longrightarrow 0\ .$$

The right hand terms constitute a partial resolution for the augmentation module $\mathbb{Z}$. We may obtain a complete Λ_μ-resolution for $\mathbb{Z}$ from the equivariant homology of $\mathbb{R}^\mu$, considered as the universal covering space of $(S^1)^\mu$. The latter space has a natural cellular structure with $\binom{\mu}{q}$ q-cells; the lifts to $\mathbb{R}^\mu$ are the Euclidean q-cubes of side 1 and whose vertices all have integral coordinates. Since $\mathbb{R}^\mu$ is contractible, there is an exact sequence

$$0 \longrightarrow C_\mu \longrightarrow C_{\mu-1} \longrightarrow \cdots \longrightarrow C_1 \longrightarrow C_o \longrightarrow \mathbb{Z} \longrightarrow 0 \qquad (3)$$

with C_q free of rank $\binom{\mu}{q}$. Alternatively, to make the maps more explicit, we may obtain the complex (C_*) as the tensor product over $\mathbb{Z}$ of μ copies of the corresponding complex for S^1 : $0 \longrightarrow \Lambda_1 \xrightarrow{t-1} \Lambda_1 \longrightarrow 0$.

All the differentials of the complex $(\mathbb{Z} \otimes_\Lambda C_*)$ are 0, so $\mathrm{Tor}_q^\Lambda(\mathbb{Z},\mathbb{Z}) = \mathbb{Z}^{\binom{\mu}{q}} = H_q((S^1)^\mu;\mathbb{Z})$. Moreover $e^q\mathbb{Z} = 0$ if $q \neq \mu$ and $e^\mu\mathbb{Z} = \mathbb{Z}$. (These are also immediate consequences of the spectral sequences of Chapter I, and of Poincaré duality and the contractability of $\mathbb{R}^\mu$). Note also that the resolution (3) together with the Crowell exact sequence for $F(\mu)$ imply that the Λ-module $F(\mu)'/F(\mu)''$ has a presentation with $\binom{\mu}{2}$ generators and $\binom{\mu}{3}$ relations.

<u>Lemma 2</u> (i) p.d. $H_2(X;\Lambda) = \max\{0, \text{p.d.} A(L) - 2\}$

(ii) <u>Either</u> p.d. $G'/G'' <$ p.d. $A(L) = \mu - 1$

<u>or</u> p.d. $A(L) <$ p.d. $G'/g'' = \mu - 2$

<u>or</u> p.d. $A(L) =$ p.d. $G'/G'' \geq \mu - 1$.

<u>Proof</u> The first assertion follows from the exact sequence (2) and Schanuel's lemma. The second assertion follows from the long exact sequence of $e^*(-) = \mathrm{Ext}_\Lambda^*(-,\Lambda)$ applied to the Crowell exact sequence in the form (1)', and the fact that $e^{q-1} I = e^q \mathbb{Z}$ for $q > 0$. //

An immediate consequence of this lemma is that if $\Delta_1(L) \neq 0$ then G'/G'' has a square presentation matrix if and only if $\mu \leq 3$. For if $\alpha(L) = 1$ then p.d. $A(L) \leq 1$ and G'/G'' is a torsion module of projective dimension $\mu - 2$ unless p.d. $A(L) \geq \mu - 1$. Since projective Λ-modules are free, a torsion module has a short projective resolution if and only if it has a square presentation matrix.

If $\mu = 1$, then $\alpha(L)$ must be 1 also and the result is well known. If $\mu = 2$, the module G'/G'' has a square presentation matrix even if $\alpha(L) = 2$. This was proven by Bailey who characterized such modules arising from 2-component links as the Λ_2-modules admitting a square presentation matrix of a particular form [7] (See also Cooper [34] and Chapter VII). It may also be seen as follows. Let $Z \subseteq D_1 = \Lambda^{n+1}$ be the submodule of 1-cycles in the cellular chain complex of X. Then there are exact sequences

$$0 \longrightarrow H_2(X;\Lambda) \longrightarrow \Lambda^n \longrightarrow Z \longrightarrow G'/G'' \longrightarrow 0$$

and

$$0 \longrightarrow Z \longrightarrow \Lambda^{n+1} \longrightarrow \Lambda \longrightarrow \mathbb{Z} \longrightarrow 0.$$

By Schanuel's lemma Z is projective, hence free, and clearly rank $Z = n$. Hence G'/G'' has a square presentation matrix. Note also that p.d. $G'/G'' \leq 2$ since $H_2(X;\Lambda)$ is either 0 (if $\alpha(L) = 1$) or free (if $\alpha(L) = 2$) by Cochran's lemma.

<u>Theorem 3</u> <u>Let G be the group of a μ-component link L with Alexander nullity α. Then</u>

(i) <u>if</u> $\mu = 1$, $E_1(L)$ <u>is principal, while if</u> $\mu \geq 1$, $E_1(L) = (\Delta_1(L)).I$;

(ii) <u>if</u> $E_{\alpha-1}(G'/G'')$ <u>and</u> $E_\alpha(L)$ <u>are both principal, or if</u> G' <u>is perfect</u> $(G' = G'')$ <u>then</u> $\mu \leq 2$;

(iii) <u>if</u> $\alpha = 1$, <u>then</u> G'/G'' <u>has no nontrivial pseudozero submodule and</u>

$\mathrm{Ann}(G'/G'') = (\Delta_1(L)/\Delta_2(L))$;

(iv) <u>for each</u> $k \geq 1$, $\alpha_k(tG'/G'')^{\sim} = (\Delta_{\alpha+k-1}(L)/\Delta_{\alpha+k}(L))$.

<u>Proof</u> (i) We may clearly assume that $E_1(L) \neq 0$, so that $A(L)$ has rank 1 and $A(L)/tA(L) \approx I$. By Theorem III.12 the isomorphism class of this ideal is just the Steinitz-Fox-Smythe row invariant of $A(L)$. From the exact sequence (2) we see that the column invariant is the class of the principal ideals, and Steinitz showed that the product of the row and column invariants was the class of the first nonzero elementary ideal [179]. The assertion follows readily.

(ii) If $E_{\alpha-1}(G'/G'')$ and $E_\alpha(L)$ are both principal, then the Crowell exact sequence gives rise to a projective resolution of $\mathbb{Z}$ of length 2

$$0 \longrightarrow (G'/G'')/(tG'/G'') \longrightarrow A(L)/(tG'/G'') \longrightarrow \Lambda \longrightarrow \mathbb{Z} \longrightarrow 0.$$

Hence $\mu = \text{p.d. } \mathbb{Z} \leq 2$. Similarly, if $G'/G'' = 0$ then $\alpha = 1$ so p.d. $A(L) \leq 1$ and $A(L) = I$ so p.d. $\mathbb{Z} \leq 2$.

(iii) If $\alpha = 1$ then p.d. $A(L) \leq 1$ and $G'/G'' = tA(L)$, so the assertions follow from Theorems III.4 and III.10 and the Remark after Theorem III.10.

(iv) This is a consequence of the Corollary to Theorem III.4.//

Part (i) of this Theorem was first proven by Torres who used properties of the Wirtinger presentation [189]. If the commutator subgroup of a 2-component link is perfect, then $\Delta_1(L) = 1$, so the linking number is ± 1, by the second Torres condition. (See Chapter VII. That the linking number is $\pm\varepsilon(\Delta_1(L))$ also follows from the Milnor presentation for G/G_3 of Chapter I. See also Chen [27]). In the knot theoretic case ($\mu = 1$) the results of part (iii) were first obtained by Crowell [39] (Note that a Λ_1-module is pseudozero if and only if it is finite). In [40] he showed that $\Delta_1(L)$ annihilates G'/G'' (under an unnecessary further hypothesis). The knot 9_{46} [157; page 399] has $G'/G'' = (\Lambda_1/(t-2)) \oplus (\Lambda_1/(2t-1))$ and so $\alpha_2(G'/G'') = (t-2, 2t-1) = (3, t+1)$ is not principal. The argument of our next theorem is related to that of Crowell in [39]

<u>Theorem 4</u> <u>Let</u> M <u>be a finitely generated</u> Λ-<u>module of rank</u> r <u>such that</u> $E_r(M)$ <u>is principal and suppose that</u> $\varepsilon(\Delta_r(M)) = \pm 1$. <u>Then</u> M <u>is torsion free as an abelian group</u>.

<u>Proof</u> Let p be an integral prime and suppose m is an element of M such that $p.m = 0$. Then $\mathrm{Ann}(\Lambda.m)$ contains p and $\mathrm{Ann}(tM)$, and hence $E_0(tM)$ by Theorem III.1, so if $\mathrm{Ann}(\Lambda.m)^{\sim} = (\delta)$, δ divides p and $\Delta_0(tM) = \Delta_r(M)$. Since $\varepsilon(\Delta_r(M)) = \pm 1$, δ must be ± 1 and so $\Lambda.m$ is pseudozero. It now follows from Theorems III.9 and III.10 that $m = 0$. //

The condition $\varepsilon(E_0(M)) = \mathbb{Z}$ is equivalent to $\mathbb{Z} \otimes_\Lambda M = 0$. For the case $\mu = 1$, $r = 0$ and $\mathbb{Z} \otimes_\Lambda M = 0$, Crowell proved this Theorem under the additional assumption that M has a square presentation matrix [39], Levine proved that this additional assumption is equivalent to such an M being torsion free as an abelian group [119], and Weber has shown that for such an M these conditions are also equivalent to $E_0(M)$ being principal [202].

When L is a 2-component link a little more can be said about $\mathbb{Z}$-torsion in G'/G''. If $\Delta_1(L) \neq 0$ then G'/G'' has nontrivial p-torsion for p a prime integer if and only if p divides $\Delta_1(L)$, in which case p must divide the linking number $\Delta_1(L)(1,1)$ (by the second Torres condition).

For Ann(G'/G") is generated by $\Delta_1(L)/\Delta_2(L)$, which is divisible by each of the prime factors of $\Delta_1(L)$. Levine has shown that given any λ in Λ_2 such that $\lambda = \bar{\lambda}$ there is a 2-component link L with linking number 0 such that $\Delta_1(L) = \lambda(t_1 - 1)(t_2 - 1)$ [117]. Hence on taking λ to be a prime integer we see that G'/G" need not be torsion free as an abelian group. If $\Delta_1(L) = 0$ and G'/G" has $\mathbb{Z}$-torsion then $E_2(L)$ cannot be principal, by Theorem 3. However since p.d. G'/G" $\leqslant 2$, as remarked above, G'/G" contains no nontrivial finite Λ-submodule, by Remark 2 after Theorem III.11.

Link module sequences

In his work on the Λ-modules A(L) and G'/G" Crowell defined a <u>link module sequence</u> as an exact sequence

$$0 \longrightarrow B \longrightarrow A \longrightarrow I \longrightarrow 0 \tag{4}$$

of Λ-modules such that A has a presentation with $n+1$ generators and n relations (for some n) and where I is the augmentation ideal of Λ. The arguments of Theorem 3 apply in this slightly more general setting, showing for instance that $E_1(A) = (\Delta_1(A))I$. In [40] Crowell showed that

(i) The first elementary ideal E_1 of A annihilates B. If $E_1 \neq 0$, then B is the torsion submodule of A;

(ii) If the product $(t_1 - 1) \ldots (t_\mu - 1)$ does not divide the Alexander polynomial Δ_1 of A, then B is annihilated by Δ_1; while if also $\mathbb{Z} \otimes_\Lambda A = \mathbb{Z}^\mu$ (as is the case with the Crowell exact sequence (1)' of a link)

(iii) The sequence (4) never splits if $\mu \geqslant 3$;

(iv) If $\mu = 2$ the sequence (4) splits if and only if $\varepsilon(\Delta_1(A)) = \pm 1$.

The results (i) and (ii) are contained in part (iii) of Theorem 3, while the other results follow on tensoring the sequence (4) over Λ with $\mathbb{Z}$. For the exact sequence can only split if $\mathbb{Z} \otimes B = 0$, as $\mathbb{Z} \otimes A = \mathbb{Z}^{\mu} = \mathbb{Z} \otimes I$. Therefore we may assume $\Delta_1(A) \neq 0$, since otherwise $E_0(\mathbb{Z} \otimes B)$ would be 0. From this and the other assumptions on A it follows that $\mathrm{Tor}_1^{\Lambda}(\mathbb{Z}, A) = \mathbb{Z}^{\mu-1}$. The long exact sequence of $\mathrm{Tor}_*^{\Lambda}(\mathbb{Z}, -)$ applied to (4) then shows that

$$\mathrm{rank}\ \mathbb{Z} \otimes B \geq \mathrm{rank}\ \mathrm{Tor}_1^{\Lambda}(\mathbb{Z}, I) - \mathrm{rank}\ \mathrm{Tor}_1^{\Lambda}(\mathbb{Z}, A) = \binom{\mu}{2} - \mu + 1$$

which is greater than 0 if $\mu \geq 3$. If $\mu = 2$ the module $\mathbb{Z} \otimes B$ has a square presentation matrix with determinant generating $\varepsilon(E_0(B))$ and so divisible by $\varepsilon(\Delta_0(B)) = \varepsilon(\Delta_1(A))$, and thus $\mathbb{Z} \otimes B = 0$ only if $\varepsilon(\Delta_1(A)) = \pm 1$.

Crowell showed also that (if $\mu = 2$ and $\Delta_1(A) \neq 0$) $\mathrm{Ext}_{\Lambda}(B, I) \approx \mathbb{Z}/\varepsilon(\Delta_1(A))$ and asked whether the class of the extension could be used to distinguish between two links.

That $\mathbb{Z} \otimes B \neq 0$ whenever $\mu \geq 3$ follows also from a result of Crowell and Strauss who showed that for any link module sequence $E_0(B) = (\Delta_1(A)).I^{\binom{\mu-2}{2}}$ [44]. This was rediscovered by Bailey [7] and extended by Traldi [190] who showed that

(a) $E_k(A) \supseteq E_{k-1}(B).I^{\mu-1}$ for all k;

(b) $E_{k-1}(B) \supseteq E_k(A).I^{\binom{\mu-1}{2}}$ for all k;

(b') $E_{k-1}(B) \supseteq E_k(A).I^{\binom{\mu-1}{2}+k-\mu}$ for $1 \leq k \leq \mu$.

(Here if $p < 0$, $E_k(A).I^p \subseteq E_{k-1}(B)$ means $E_k(A) \subseteq E_{k-1}(B).I^{-p}$). Note that (b') implies $E_0(B) \supseteq (\Delta_1(A)).I^{\binom{\mu-2}{2}}$, since $E_1(A) = (\Delta_1(A)).I$, thus proving part of the Crowell-Strauss result. We shall sketch a proof of (a) and of part of (b).

On applying lemma III.6 to the link module sequence (4) we see that $E_k(A) \supseteq E_{k-1}(B).E_1(I)$ for any k, and it is not hard to show by induction on μ that $E_1(I) = I^{\mu-1}$, using the presentation $C_2 \to C_1 \to I \to 0$ derived from the Λ-resolution for $\mathbb{Z}$ given above. (This is also the Jacobian presentation for $I = A(\mathbb{Z}^{\mu})$ obtained via the free differential calculus from an obvious presentation for the group $\mathbb{Z}^{\mu}$. See also Lemma 5.2 of [44].) This proves (a).

The link module sequence (4) together with the Crowell sequence for $F(\mu)$ gives rise to another short exact sequence

$$0 \longrightarrow F(\mu)'/F(\mu)'' \longrightarrow B \oplus \Lambda^{\mu} \longrightarrow A \longrightarrow 0.$$

Hence $E_{k-1}(B) = E_{k+\mu-1}(B \oplus \Lambda^{\mu}) \supseteq E_k(A).E_{\mu-1}(F(\mu)'/F(\mu)'')$ for any $k \geq 1$. Thus (b) is true in general if and only if it is true for the Crowell exact sequence for $F(\mu)$ when $k = \mu$. (This special case is established in Lemma 5.6 of [44].)

Completion of link module sequences

Stallings' theorem implies that the nilpotent quotients $G/G''G_n = (G/G_n)/(G/G_n)''$ of a link group are invariant under (possibly wild) I-equivalence of the link. These quotients have been called the Chen groups of the link by Murasugi, who used free differential calculus to show that, if $\mu = 2$, these groups are "free" if and only if $E_{\mu-1}(L) = 0$ if and only if the longitudes of L are in $G(\infty) = \bigcap_{n \geq 2}(G_nG'')$ [140]. In the next chapter we shall give a new proof of this result, applicable to links with any number of components, and we observe that the Alexander nullity of a link is invariant under I-equivalence. Massey has also extended the first equivalence of Murasugi's theorem, using commutative algebra in a similar but more whole-hearted way than we do [125]. Although he obtains other interesting results, our mixture of commutative algebra and group theory seems necessary to derive the condition on the longitudes, and so we shall only sketch proofs of some of his results here.

He observed that the nilpotent completion of G/G'' corresponds to the I-adic completion of $B = G'/G''$, since $B/I^nB = G'/G''G_{n+2}$, and so considered the I-adic completion of a link module sequence (4) such that $\mathbb{Z} \otimes A = \mathbb{Z}^\mu$. (Since completion is an exact functor, the completed sequence is also exact.) Let $\hat{M}$ denote the I-adic completion of a Λ-module M, so $\hat{M}$ is a $\hat{\Lambda}$-module, and Λ embeds in $\hat{\Lambda} \approx \mathbb{Z}[[X_1,\ldots,X_\mu]]$ via $t_i \rightarrowtail 1 + X_i$. Then he proved the following theorems. (We have changed the notation and abbreviated his enunciation slightly.)

I The $\hat{\Lambda}$-module $\hat{A}$ has a presentation with μ generators and $s < \mu$ relations.

II The $\hat{\Lambda}$-module $\hat{B}$ has a presentation with $\binom{\mu}{2}$ generators and $\binom{\mu}{3} + s$ relations. Moreover $\binom{\mu}{3}$ of these relations are the same for all μ-component links.

III If $\mu = 2$ the associated graded module $G(B) = G(\hat{B})$ is a cyclic module over $G(\Lambda) = G(\hat{\Lambda}) = \mathbb{Z}[X_1,\ldots,X_n]$, and Ann $G(B)$ is generated by the "initial form" of the image of the Alexander polynomial $\Delta_1(A)$ in $\hat{\Lambda}$. Thus the Chen groups of a 2-component link are effectively determined by its Alexander polynomial.

(Here the <u>initial form</u> of a power series of $\hat{\Lambda}$ is the homogeneous polynomial in X_i consisting of the nonzero terms of lowest degree.)

IV The completed Crowell exact sequence

$$0 \longrightarrow G'/G'' \longrightarrow A(L) \longrightarrow \hat{I} \longrightarrow 0$$

of a link is invariant under I-equivalence (hence under isotopy and concordance).

Corollary The principal ideal in the power series ring $\hat{\Lambda}$ generated by the image of the Alexander polynomial is an invariant of the link under I-equivalence.

The first two theorems follow from the link module sequence and the standard presentation of I, on using Nakayama's lemma, and the third is a consequence of the second, together with a little group theory. Massey observes that if $\ell = \Delta_1(L)(1,1) \neq 0$ then the initial form of $\Delta_1(L)$ is the constant ℓ, while if $\ell = 0$ the Torres conditions (see Chapter VII) imply

(a) the initial form of $\Delta_1(L)$ is an homogeneous polynomial of even degree in X_1 and X_2;

(b) if the initial form has degree n, the coefficients of X_1^n and X_2^n are both 0.

He asks whether these characterize such initial forms, and verifies that they do for $n = 2$, and that any even degree can occur. The fourth theorem is a consequence of Stallings' theorem, while the corollary follows from the fact that the principal ideal generated by the image of $\Delta_1(L)$ in $\hat{\Lambda}$ is the ideal $(\Delta_1(\hat{A}(L)))$. This corollary may be restated in the following form, derived earlier by Kawauchi [89].

Corollary Let $\Delta_1(L) = \delta_1.u_1$ where $\varepsilon(u_1) = \pm 1$ and δ_1 has no factor augmenting to ± 1. Then δ_1 is invariant under I-equivalence.

For an element of Λ becomes a unit in $\hat{\Lambda}$ if and only if it augments to ± 1. (Kawauchi's argument applies only to PL I-equivalences, as it assumes that the equivariant chain complexes of the maximal abelian cover of the complement of the I-equivalence are finitely generated.)

We may now give a simple proof of the following theorem.

Theorem 5. Let $L: \mu S^1 \to S^3$ be a link with group G. If G is nilpotent, then $G = 1, \mathbb{Z}$ or $\mathbb{Z}^2$.

Proof. If L has nilpotent group then so do all its sublinks. So we may assume that $\mu \leq 3$. It will suffice to prove that $G_2 = G_3$ and that $\mu \leq 2$. Since these are true if $\mu \leq 1$, we may assume that $\mu \geq 2$. Let $B = G'/G''$. Since $G' = G_2 \supseteq G_3 \supseteq G''$, it will suffice to prove that $B = IB$, i.e. that $\hat{B} = 0$. As G is nilpotent, B is finitely generated as an abelian group, and $I^n B = 0$ for n large, so $B = \hat{B}$. Massey's Theorem II implies that $\hat{B}$ has deficiency ≥ 0 as a $\hat{\Lambda}$-module, and so can only be finitely generated as an abelian group if it is 0. Therefore G is abelian and so isomorphic to $\mathbb{Z}^\mu$. Since $H_2(G;\mathbb{Z}) = \mathbb{Z}^{\binom{\mu}{2}}$ is a quotient of $H_2(X;\mathbb{Z}) = \mathbb{Z}^{\mu-1}$ [83], $\mu \leq 2$. //

(Using the Sphere Theorem and the Loop Theorem one can in fact show that if a link has solvable group then the link is empty, the unknot or  .)

CHAPTER V THE VANISHING OF ALEXANDER IDEALS

At the 1961 Georgia conference on Topology of 3-Manifolds, Fox raised the question of the geometric significance of the identical vanishing of the first Alexander polynomial of a 2-component link [50; Problem 16]. Boundary links clearly satisfy this condition, but in 1965 Smythe introduced the concept of "homology boundary link" to show that such a link need not be boundary [174]. He conjectured in turn that "$\Delta_1(L) = 0$" should imply that L be an homology boundary link. In 1970 Murasugi proved that this condition is equivalent to "each of the subquotients $G''G_n/G''G_{n+1}$ is isomorphic to the corresponding subquotient of F(2) " and to "the longitudes of L are in $G(\infty) = \bigcap_{n\geq 2} (G''G_n)$ " [140]. Cochran showed in his 1970 Dartmouth thesis that "$\Delta_1(L) = 0$" implied that $H_2(X;\Lambda) = \Lambda$, and constructed a family of unsplittable 2-component links with first Alexander polynomial 0 [29, 30].

In this chapter is given a counter-example to the conjecture of Smythe, as an illustration of a new criterion for a ribbon link to be an homology boundary link. The more general situation of the vanishing of certain of the Alexander ideals of a finitely generated group with abelianization $\mathbb{Z}^\mu$ is considered, and it is shown that the rank of the Alexander module A(G) depends only on the nilpotent quotients of G/G". As a consequence the Alexander nullity of a link is an invariant of arbitrary I-equivalence. Conversely if $\alpha(G) = \mu$ then any map of $F(\mu)$ to G inducing an isomorphism on abelianization induces isomorphisms on all such nilpotent quotients. Furthermore if G is the group of a μ-component link L, then $\alpha(L) = \mu$ if and only if the longitudes of L lie in $G(\infty)$. We shall also answer a question raised by Cochran (for 2-component links), by showing that $H_2(X;\Lambda) = H_2(X';\mathbb{Z})$ projects onto $H_2(X;\mathbb{Z})$ if and only if $E_{\mu-1}(L) = 0$.

The Counter-example

An epimorphism of groups $f:G \to H$ induces an epimorphism $A(f):A(G) \to A(H)$. Therefore if G is finitely generated, so the elementary ideals $E_*(G) = E_*(A(G)) \subseteq \mathbb{Z}[G/G']$ are defined, the image of $E_i(G)$ in $\mathbb{Z}[H/H']$ is contained in $E_i(H)$. Now let H be a group with a presentation of deficiency μ and with abelianization $H/H' = \mathbb{Z}^\mu$. Then $E_i(H) = 0$ for $i < \mu$ and $E_\mu(H) \equiv (1)$ modulo I. Hence if L is an homology boundary link, so $G(L)$ maps onto $H = F(\mu)$, then $E_{\mu-1}(L) = 0$. By Theorem II.6 this is also true of ribbon links (taking $H = H(R)$) and so they provide examples on which to test Smythe's conjecture. Since any epimorphism $G(L) \to F(\mu)$ induces an isomorphism $G/G_\omega \approx F(\mu)$, it must factor through $H(R)$ if L bounds a ribbon $R:\mu D^2 \to S^3$. Thus the criterion of the next theorem may suffice to show that a ribbon link is not an homology boundary link.

<u>Theorem 1</u> <u>If</u> H <u>is a group with a presentation of deficiency</u> μ <u>and with</u> $H/H' = \mathbb{Z}^\mu$, <u>which maps onto</u> $F(\mu)/F(\mu)''$, <u>then</u> $E_\mu(H)$ <u>is principal.</u>

<u>Proof</u> The assumptions imply that $A(H)$ has rank μ and maps onto Λ^μ. Therefore $A(H) \approx \Lambda^\mu \oplus tA(H)$. Since $A(H)$ has a presentation of deficiency μ, adding μ relations to kill a basis for the free summand gives a square presentation matrix for $tA(H)$. Therefore $E_\mu(H) = E_\mu(A(H)) = E_0(tA(H))$ is principal. //

In the next chapter we shall give several partial converses of this theorem. We shall now present our first counter-example to Smythe's conjecture.

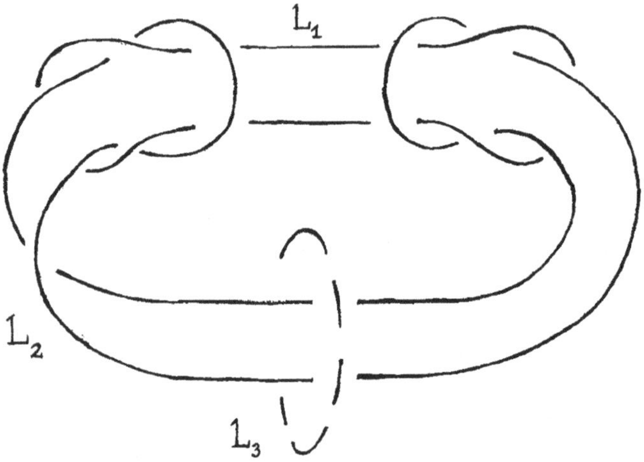

Figure 1

The solid link in Figure 1 (which has unknotted components) extends to a ribbon map R with 4 singularities. The ribbon group H(R) for this ribbon has a presentation

$$\{x_1,x_2,x_3,y_1,y_2,y_3 \mid y_1^{-1}x_1y_1 = x_2, y_3x_2y_3^{-1} = x_3, x_1^{-1}y_1x_1 = y_2, x_3^{-1}y_2x_3 = y_3\}$$

which is Tietze-equivalent [43; page 43] to

$$\{x_1,x_3,y_1 \mid x_1y_1^{-1}x_1y_1x_1^{-1} = (x_3x_1^{-1})^{-1}x_1^{-1}y_1^{-1}x_1(x_3x_1^{-1})y_1x_1(x_3x_1^{-1})\}$$

so H(R) has a preabelian presentation

$$\{x,\ y,\ a \mid xy^{-1}\ xyx^{-1} = a^{-1}\ x^{-1}\ y^{-1}\ x\ a\ y\ x\ a\}$$

The Jacobian matrix of this presentation is

$$M = \|\ (y-1)(x^{-1}y^{-1} - xy^{-1}), (1-x)(x^{-1}\ y^{-1} - xy^{-1}),\ 1 - y^{-1} - x\ \|$$

ans so $E_2(H(R)) = ((y-1)(x^2-1),\ (1-x)(x^2-1),\ y - 1 - xy)$

$= (x+1,\ y-1-xy) \qquad \text{since } x-1 = y^{-1}(1-(y-1-xy))$

$= (x+1,\ 2y-1)$

which is clearly not principal. Thus H(R) cannot map onto F(2)/F(2)"; a fortiori, the group of the link cannot map onto F(2).

We shall give several other proofs that this link is not an homology boundary link in Chapters VI and VIII. (Note however that on removing the half twist from the lower ribbon we obtain Milnor's boundary link [130; page 305]). In Chapter VI we shall also give an example of a ribbon link which is not an homology boundary link although $G/G'' \approx F(2)/F(2)''$, so that this cannot be proven using Alexander ideals.

The link of Figure II.1 is a ribbon homology boundary link for which $E_2(L) = (x-1,\ y^2-y+1)$, so $E_\mu(H(R))$ principal need not imply $E_\mu(L)$ principal. This is the simplest nonsplittable link with Alexander polynomial 0, as may be seen from the tables in [157]. (Smythe's original homology boundary link may be obtained by giving the knotted ribbon of this link 3 half twists). It is not hard to verify that any ribbon counter example to Smythe's conjecture must have at least 4 ribbon singularities. The examples constructed by means of Baumslag's parafree group in Chapter II show that the higher dimensional analogue of Smythe's conjecture is false. It is clear that for these groups $E_1(G) = 0$, but more generally $E_{\mu-1}(G) = 0$ for G the group of any μ-component n-link (for $n \geq 2$) as follows from Stallings' theorem and the theorem of the next section.

Alexander Ideals and Chen Groups

In [140] Murasugi proved, *inter alia*, that for G the group of a 2-component link the following conditions are equivalent

(1) $E_1(G) = 0$;

(2) the Chen group $Q(G;q) = G_qG''/G_{q+1}G''$ is isomorphic to $Q(F(2);q)$ for all $q \geq 1$;

(3) the longitudes of the link are in $G(\infty) = \bigcap_{q \geq 1} (G_qG'')$.

In the course of his proof, which involved delicate computations in the free differential calculus, he found presentations for the (finitely generated abelian) groups $Q(H;q)$ for H free of finite rank and for H the group of a 2-component link.

In this section we shall give a proof of the following generalization (which was published in [72]).

Theorem 2 *If G is a finitely generated group with $G/G' \approx \mathbb{Z}^{\mu}$ then the following two conditions are equivalent:*

(1) $E_{\mu-1}(G) = 0$;

(2) $Q(G;q) \approx Q(F(\mu);q)$ *for all integers* $q \geq 1$.

Furthermore, if G is the group of a μ-component link L then (1) and (2) are equivalent to

(3) *the longitudes of L are in* $G(\infty)$.

Instead of calculating free derivatives, we shall use Nakayama's lemma and Krull's theorem, applied to the Crowell exact sequences for the groups G/G_n. (As indicated in Chapter IV, Massey has also used commutative algebra to extend Murasugi's equivalence (1) ⇔ (2) to arbitrary link groups, and he has shown that the Chen groups of a

2-component link may be effectively computed from the Alexander polynomial [125]).

For any group H, let $\bar{H} = H/H''$. Then H/H_qH'', $\bar{H}/(\bar{H})_q$ and $\overline{H/H_q}$ are naturally isomorphic. A map $f:H \to K$ induces maps $\bar{f}:\bar{H} \to \bar{K}$ and $f_q:H/H_q \to K/K_q$, and $(\bar{f})_q$ is naturally equivalent to $(\bar{f}_q)$. We shall identify these naturally isomorphic quotients of H and naturally equivalent maps. There are short exact sequences $1 \to Q(H;q) \to H/H_{q+1}H'' \to H/H_qH'' \to 1$ by definition of the Chen groups, and so by the five lemma and induction a map $f:H \to K$ induces isomorphisms on all Chen groups if and only if all the maps $\bar{f}_q$ are isomorphisms. The arguments below will be in terms of the groups H/H_qH'' excepting for one appeal to the computation of the Chen groups of a free group by Chen and Murasugi.

The qth truncated Alexander module of H is the $\mathbb{Z}[H/H']$-module $A_q(H) = \mathfrak{h}/\mathfrak{h}_2\mathfrak{h} + \mathfrak{h}^q$; in particular $A_2(H) = \mathfrak{h}/\mathfrak{h}^2$ is isomorphic to H/H' (see Chapter IV). Given a finite presentation for H and an isomorphism $H/H' = \mathbb{Z}^\mu$, the Jacobian matrix of this presentation at this map is a presentation matrix for A(H) over Λ_μ, and this matrix reduced modulo I^{q-1} is a presentation matrix for $A_q(H)$ over Λ/I^{q-1} [175]. Hence $A_q(F(\mu)) = (\Lambda/I^{q-1})^\mu$. Let $\mathfrak{h}_q$ be the ideal generated by $\{w-1 \mid w \in H_q\}$, for each $q \geq 1$. Then since $[x,h]-1 = ((x-1)(h-1) - (h-1)(x-1)\ x^{-1}\ h^{-1}$ it follows by induction that $\mathfrak{h}_q \subseteq \mathfrak{h}^q$ for each $q \geq 1$. Then $A(H/H_n) = \mathfrak{h}/\mathfrak{h}_2\mathfrak{h} + \mathfrak{h}_n$ and $A_q(H/H_n) = \mathfrak{h}/\mathfrak{h}_2\mathfrak{h} + \mathfrak{h}_n + \mathfrak{h}^q$ [175]. If $f:H \to K$, let A(f) and $A_q(f)$ be the maps induced on A(H) and $A_q(H)$ respectively. In particular, the quotient map $p:H \to \bar{H}$ induces isomorphisms A(p) and $A_q(p)$.

<u>Proof of the Theorem</u> Choose a map $\theta : F = F(\mu) \to G$ inducing an isomorphism $\theta_2 : \mathbb{Z}^\mu \to G/G'$, and hence an identification of $\Lambda = \Lambda_\mu$ with $\mathbb{Z}[G/G']$. (Such a map shall be referred to below as a "meridian" map). Let $\tilde{\theta} : \mathbb{Z}[F] \to \mathbb{Z}[G]$ be the induced map of group rings.

(1) implies (2). Suppose $E_{\mu-1}(G) \subset I^{q-1}$. Let $R = \Lambda / I^{q-1}$. Then $R/\mathrm{rad}\, R = R/I = \mathbb{Z}$, since the image of I in R is nilpotent and $\mathbb{Z}$ is a domain. Since $A_q(\theta) \otimes_R \mathbb{Z} = \theta_2$ is an isomorphism, the map $A_q(\theta) : R^\mu = A_q(F) \to A_q(G)$ is onto by Nakayama's lemma [4; page 21]. The kernel of this map is finitely generated, since R is noetherian, and so $A_q(G)$ has a presentation

$$R^a \xrightarrow{M} R \xrightarrow{A_q(\theta)} A_q(G) \to 0.$$

Since $E_{\mu-1}(A_q(G)) = E_{\mu-1}(G)$ reduced modulo $I^{q-1} = 0$, the matrix M must be null, and so $A_q(\theta)$ is an isomorphism. Thus if $E_{\mu-1}(G) = 0$, the maps $A_q(\theta)$ are isomorphisms for every q.

By the Crowell equivalence of Chapter IV, to show that the maps $\bar{\theta}_q$ are isomorphisms (and hence that the Chen groups are isomorphic), it will suffice to show that each $A(\theta_q) : A(F/F_q) \to A(G/G_q)$ is an isomorphism. Since θ_2 is onto, $\theta_q : F/F_q \to G/G_q$ is onto [123; page 350] and so $A(\theta_q)$ is onto. On considering (for each r) the commutative diagram

$$\begin{array}{ccc} A(F/F_q) & \longrightarrow & A_r(F/F_q) \\ \downarrow & & \downarrow \\ A(G/G_q) & \longrightarrow & A_r(G/G_q) \end{array}$$

in which each map is onto, it will suffice to show that the map $A_r(F/F_q) \to A_r(G/G_q)$ is a monomorphism for each r, and that $\bigcap_{r \geq 1} \ker(:A(F/F_q) \to A_r(F/F_q)) = 0$. The map

$A_r(\theta): \mathcal{F}/\mathcal{F}_2\mathcal{F} + \mathcal{F}^r \longrightarrow \mathcal{G}/\mathcal{G}_2\mathcal{G} + \mathcal{G}^r$ is an isomorphism, so $\ker \tilde{\theta} \subseteq \tilde{\theta}^{-1}(\mathcal{G}_2\mathcal{G} + \mathcal{G}^r) = \mathcal{F}_2\mathcal{F} + \mathcal{F}^r$. Clearly $\tilde{\theta}(\mathcal{F}_2\mathcal{F} + \mathcal{F}^r + \mathcal{F}_q) \subseteq \mathcal{G}_2\mathcal{G} + \mathcal{G}^r + \mathcal{G}_q$, and since $\theta: F \to G$ is onto modulo G_r, it follows that $\mathcal{G}_s \subseteq \tilde{\theta}(\mathcal{F}_s) + \mathcal{G}^r$, for each s. (For if g_s is in G_s, then $g_s = \theta(f_s)g_r$ for some f_s in F_s and g_r in G_r by induction on s, and hence $g_s - 1 = \theta(f_s)g_r - 1 = \theta(f_s) - 1 + \theta(f_s)(f_r - 1)$ is in $\tilde{\theta}(\mathcal{F}_s) + \mathcal{G}_r \subseteq \tilde{\theta}(\mathcal{F}_s) + \mathcal{G}^r$). Therefore $\mathcal{G}_2\mathcal{G} + \mathcal{G}^r + \mathcal{G}_q \subseteq \mathcal{G}_2\mathcal{G} + \mathcal{G}^r + \tilde{\theta}(\mathcal{F}_q)$ and so $\tilde{\theta}^{-1}(\mathcal{G}_2\mathcal{G} + \mathcal{G}^r + \mathcal{G}_q) = \tilde{\theta}^{-1}(\mathcal{G}_2\mathcal{G} + \mathcal{G}^r) + \mathcal{F}_q = \mathcal{F}_2\mathcal{F} + \mathcal{F}^r + \mathcal{F}_q$, that is, the map from $A_r(F/F_q) = \mathcal{F}/\mathcal{F}_2\mathcal{F} + \mathcal{F}^r + \mathcal{F}_q$ to $A_r(G/G_q) = \mathcal{G}/\mathcal{G}_2\mathcal{G} + \mathcal{G}^r + \mathcal{G}_q$ is a monomorphism (and hence an isomorphism). Now $A(F) = \Lambda^\mu$ and $A_r(F) = (\Lambda/I^{r-1})^\mu = A(F)/I^{r-1}A(F)$, so on considering the commutative diagram

$$\begin{array}{ccc}
A(F) & \longrightarrow & A(F/F_q) \\
\downarrow & & \downarrow \\
A_r(F) & \longrightarrow & A_r(F/F_q)
\end{array}$$

it follows that $\ker(: A(F/F_q) \to A_r(F/F_q)) = I^{r-1}A(F/F_q)$. Since $A(F/F_q)$ is finitely generated over the noetherian ring Λ, $\bigcap_{r \geq 1}(I^{r-1}A(F/F_q)) = \{\alpha \text{ in } A(F/F_q) \mid (1+j)\alpha = 0 \text{ for some } j \text{ in } I\}$ by Krull's theorem [4; page 110].

Now $A(F/F_q)$ sits in the Crowell sequence

$$0 \to F'/F_qF'' \to A(F/F_q) \to \Lambda \to \mathbb{Z} \to 0$$

so if α in $A(F/F_q)$ is such that $(1+j)\alpha = 0$ for some j in I, then α is in F'/F_qF''. But $F'/F_qF'' = (F'/F''/I^{q-1}(F'/F''))$ is a module over Λ/I^{q-1} and $1+j$ is invertible in Λ/I^{q-1}, for any j in I. Therefore $\alpha = 0$. This completes the argument for (1) $\Rightarrow$ (2).

(2) implies (1). Suppose that there is an homomorphism $\psi: G \to H$ which

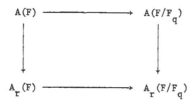

induces an isomorphism $\bar{\psi}_q : \bar{G}/\bar{G}_q \to \bar{H}/\bar{H}_q$. Then the induced maps $A_r(\bar{G}/\bar{G}_q) \to A_r(\bar{H}/\bar{H}_q)$ are isomorphisms for all r. Since $A_r(K/K_q) = \mathfrak{k}/\mathfrak{k}_2\mathfrak{k} + \mathfrak{k}^r + \mathfrak{k}_q$ and $\mathfrak{k}_q \subseteq \mathfrak{k}^q \subseteq \mathfrak{k}^r$ for $r \leq q$, $A_r(K/K_q) = A_r(K)$ if $r \leq q$. It follows that the induced map $A_q(\bar{G}) \to A_q(\bar{H})$ is an isomorphism and so $E_*(G) \equiv E_*(H)$ modulo I^{q-1}. Hence if ψ induces isomorphisms on all Chen groups, so all the maps $\bar{\psi}_q$ are isomorphisms (as above) the $E_j(G) = 0$ if and only if $E_j(H) = 0$ (since $\bigcap_{q \geq 1} I^{q-1} = 0$). In particular, if the Chen groups of G are "free", then the maps $\bar{\theta}_q$ induced by a "meridian" map θ, which are always epimorphisms, are isomorphisms by induction, the five lemma and the hopficity [123; page 296] of finitely generated abelian groups $Q(F;q)$ (applied to the commutative diagrams

$$\begin{array}{ccccccccc} 1 & \longrightarrow & Q(F;q) & \longrightarrow & F/F_{q+1}F'' & \longrightarrow & F/F_qF'' & \longrightarrow & 1 \\ & & \downarrow & & \downarrow \bar{\theta}_{q+1} & & \downarrow \bar{\theta}_q & & \\ 1 & \longrightarrow & Q(G;q) & \longrightarrow & G/G_{q+1}G'' & \longrightarrow & G/G_qG'' & \longrightarrow & 1 \end{array})$$

and so $E_{\mu-1}(G) = 0$

(3) implies (2). Suppose now that G is the group of a link L. Then by Milnor's theorem G/G_q has a presentation $\{x_i,\ 1 \leq i \leq \mu \mid [x_i, e_i]\ 1 \leq i \leq \mu$, commutators of weight $q+1\}$ where e_j is a word representing the image of the jth longitude in G/G_q, so if the longitudes are all in $G(\infty)$, G/G_qG'' is "free", and a "meridian" map induces isomorphisms of $Q(F;q)$ with $Q(G:q)$

(2) implies (3)

Lemma Let $f_1,\ldots,f_\mu$ generate F/F_qF''. Then the centraliser of f_μ in F/F_qF'' is generated by f_μ and $F_{q-1}F''/F_qF''$.

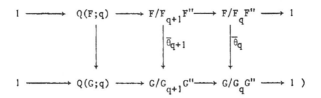

Proof Suppose $f_\mu c = c f_\mu$. Let $\lambda_i : F/F_qF'' \to \mathbb{Z}$ be defined by $\lambda_i(f_j) = \delta_{ij}$ for $1 \leq i, j \leq \mu$. Without loss of generality, it may be assumed that $\lambda_\mu(c) = 0$. To show that $\lambda_j(c) = 0$ for all j it suffices to pass to the quotient group obtained by killing f_i for all $i \neq j,\mu$ and then to $F(2)/F(2)_3$ where it is clear. So it may be assumed that c is in $(F/F_qF'')'$. Now $[-,f_\mu]$ induces a 1-1 map $Q(F;r) \to Q(F;r+1)$ for $r \geq 2$ (it maps distinct standard elements of length r to distinct standard elements of length r + 1 [25,140]) and so by induction c is in $F_{q-1}F''/F_qF''$ and the lemma is proven.

Consequently if the Chen groups of G are "free", then the quotients G/G_qG'' are "free", and since they are generated by the meridians, and since the jth longitude commutes with the jth meridian, it follows by induction that all the longitudes are in $G(\infty)$. //

The image of the Chen kernel $G(\infty)$ in G'/G'' always lies in tG'/G'', since $G(\infty)/G'' = \bigcap_{n\geq 2}(G_nG''/G'') = \bigcap_{n\geq 2} I^{n-2}(G'/G'') = \{g \text{ in } G'/G'' \mid (1+j)g = 0$ for some j in $I\}$. If $E_{\mu-1}(L) = 0$, $G(\infty)$ maps onto tG'/G''. For $E_\mu(L)$ contains some δ such that $\varepsilon(\delta) = 1$, since $\varepsilon(E_\mu(L)) = \mathbb{Z}$. Let $S = \{\delta^n \mid n \geq 0\}$. Then $A(L)_S$ is a projective Λ_S-module by the Corollary to Theorem III.9, so some power δ^N of δ annihilates $tA(L)$. Hence tG'/G'' is annihilated by $1 + j$, where $j = \delta^N - 1$ is in I. Conversely if $G(\infty)/G'' = tG'/G''$ and if the linking numbers of L are all 0, so that the longitudes are in G' and hence have image in tG'/G'', then they are in $G(\infty)$ and $E_{\mu-1}(L) = 0$.

If G has a presentation of deficiency μ and $G/G' = \mathbb{Z}^\mu$ then $E_{\mu-1}(G) = 0$ so the Chen groups are "free". In fact the nilpotent quotients G/G_q are then also "free" [123; page 353]. For link groups the nilpotent quotients G/G_q are all "free" if and only if the longitudes are in G_ω, as may be seen by arguments similar to those for (2) $\Leftrightarrow$ (3). In particular, this is the case if the link is I-equivalent to an homology boundary link, for the nilpotent quotients G/G_q are invariant under I-equivalence, by Stallings' theorem. If G is a link group with all Chen groups "free", are all the nilpotent quotients G/G_q "free"? This is certainly false for other groups, for instance the group presented by $\{x,y \mid [[x,y],[x,y^{-1}]]\}$.

(Is this a link group?) More geometrically, does the vanishing of $E_{\mu-1}(G)$ for G the group of a μ-component link L imply that L is concordant to a boundary link, or at least I-equivalent to an homology boundary link?

The argument used to show that (2) implies (1), together with the invariance of the nilpotent quotients under I-equivalence, actually gives the stronger

Corollary The Alexander nullity of a link is invariant under I-equivalence. In particular if L is a μ-component slice link then $\alpha(L) = \mu$.

This corollary is also a consequence of Massey's Theorem IV (see Chapter IV) and has also been found independently by Kawauchi [90] and Sato [165]. (However their proofs apply only for PL I-equivalences).

A Question of Cochran

A 2-component boundary link L may also be characterized as one for which there is a connected closed surface C in S^3 which separates the components of L and such that each component is nullhomologous in the complement of C. Such a surface represents a generator of $H_2(X;\mathbb{Z})$ and lifts to a generator of $H_2(X';\mathbb{Z}) = H_2(X;\Lambda)$. In attempting to decide whether 2-component links with first Alexander polynomial 0 were homology boundary links, Cochran showed that for such links $H_2(X;\Lambda)$ is a free module of rank 1, and asked whether the map to $H_2(X;\mathbb{Z})$ induced by the projection p was onto [29] (See Chapter IV). It is clear that this is only possible if $E_1(L) = 0$. In this short section we shall show that this condition is also sufficient, thereby answering Cochran's question affirmatively, and our argument shall resolve the corresponding question for links with more than 2 components.

Theorem 3 The cokernel of the natural map $p_2 : H_2(X;\Lambda) \to H_2(X;\mathbb{Z})$ is $\mathrm{Tor}_1^{\Lambda}(\mathbb{Z}, A(L))$, and is invariant under I-equivalence. The map p_2 is onto if and only if $E_{\mu-1}(L) = 0$.

Proof The first assertion follows from the Cartan-Leray spectral sequence for the projection $p:(X', p^{-1}(*)) \to (X,*)$ which gives rise to an isomorphism $\mathbb{Z} \otimes_{\Lambda} A(L) \approx \mathbb{Z}^{\mu}$ and to an exact sequence

$$0 \to \mathrm{Tor}_2^{\Lambda}(\mathbb{Z},A(L)) \to \mathbb{Z} \otimes_{\Lambda} H_2(X;\Lambda) \to H_2(X;\mathbb{Z}) \to \mathrm{Tor}_1^{\Lambda}(\mathbb{Z},A(L)) \to 0.$$

Let $\hat{M}$ denote the I-adic completion of a Λ-module M and let $T = \mathrm{Tor}_1^{\Lambda}(\mathbb{Z},A(L))$. Then $\mathbb{Z} = \hat{\mathbb{Z}}$ and $T = \hat{T}$ since $I\,\mathbb{Z} = I\,T = 0$. Since the completion of an exact sequence of finitely generated Λ-modules is exact, $\hat{T} = \mathrm{Tor}_1^{\hat{\Lambda}}(\mathbb{Z},\hat{A}(L))$ [13 ; page 203]. The second assertion now follows, as $\hat{A}(L)$ is invariant under I-equivalence, by Theorem IV of Massey. (See also Theorem 2).

Now let R be the localization Λ_I, and let $A = A(L)_I$. Then $R/IR = \mathbb{Z}_I = \mathbb{Q}$, and $A/IA = \mathbb{Z}_I^{\mu} = \mathbb{Q}^{\mu}$, so by Nakayama's lemma there is an epimorphism $\phi : R^{\mu} \twoheadrightarrow A$. If p_2 is onto then $\mathrm{Tor}_1^R(\mathbb{Q}, A) = T_I = 0$, so $\mathbb{Q} \otimes_R \ker \phi = 0$ and by Nakayama' lemma ϕ is an isomorphism. (Cf. [13; page 84]). Hence $\alpha(L) = \Lambda\text{-rank } A(L) = R\text{-rank } A = \mu$, so $E_{\mu-1}(L) = 0$. Conversely, if $E_{\mu-1}(L) = 0$ then $\hat{A}(L)$ is a free $\hat{\Lambda}$-module of rank μ, by Theorem I of Massey, so $T = \hat{T} = 0$ and p_2 is onto. //

If L is a 2-component homology boundary link then there is a map $f:X \to S^1 \vee S^1$ inducing an epimorphism $f_*:G \to F(2)$. Does the inverse image of the wedge point serve as a singular separating surface for L? In general is there a geometrically significant generator for $H_2(X;\Lambda)$ which projects nicely?

CHAPTER VI LONGITUDES AND PRINCIPALITY

In contrast to the situation discussed in the previous chapter, the μth Alexander ideal of a μ-component link never vanishes. Indeed it is necessarily comaximal with the augmentation ideal I, for evaluating the Jacobian matrix at $(1,\ldots,1)$ gives a presentation for $\mathbb{Z}^{\mu}$, the abelianization of the link group. In this chapter we shall be concerned with links for which $E_{\mu}(L)$ is the first nonzero ideal and in particular when (for such links) this ideal is principal. We shall relate the latter condition to the condition that the longitudes lie in G''. These conditions were separately hypothesized as characterizations of boundary links [174], but we shall give an example to show that they are not sufficient. We shall give necessary and sufficient conditions for the Alexander module $A(G)$ to map onto $\Lambda^{\mu} = A(F(\mu)) = A(F(\mu)/F(\mu)'')$. In the 2-component case an equivalent condition is that G maps onto $F(2)/F(2)''$ and in this case we can show

$$E_2(L) = (\Delta_2(L))(b_1(t_2) + b_2(t_1) - 1, (t_1 - 1)b_2(t_1), (t_2 - 1)b_1(t_2))$$

for some $b_1(t_2)$, $b_2(t_1)$ in Δ_2 such that $b_1(1) = b_2(1) = 1$. This result, and the relationship between principality and longitudes were first announced by Crowell and Brown, in the case of 2-component homology boundary links, but no proof has yet been published. In a letter to Smythe (dated 20 May 1976, [42]), Crowell stated the following results:

"Let $L = \ell_1 \cup \ell_2$ be a 2-component homology boundary link with group $G = \pi_1(S^3 - L)$. Let $s,t \in G/G'$ be the classes of the meridians. Then $\mathbb{Z}[G/G'] = \mathbb{Z}[s,s^{-1},t,t^{-1}]$.

<u>Theorem I</u> There are polynomials $b(s) \in \mathbb{Z}[s]$, $c(t) \in \mathbb{Z}[t]$ such that $b(1) = c(1) = 1$ and $E_2(L) = (\Delta_2(L)).I$ where $I = ((s-1)b(s), (t-1)c(t), b(s) + c(t) - 1)$.

Algebraic properties of the ideal I:

1. g.c.d. $((s-1)b(s),(t-1)c(t),b(s)+c(t)-1) = 1$;

2. I principal $\Leftrightarrow I = (1) \Leftrightarrow b(s)$, $c(t)$ units in $\mathbb{Z}[G/G']$;

3. I determines $b(s)$, $c(t)$;

4. E_2 determines I (since it determines Δ_2, and $I = (E_2 : (\Delta_2))$).

Theorem II If L is a boundary link, then b(s) and c(t) are units in $\mathbb{Z}[G/G']$.

Corollary of Theorems I and II If L is a boundary link, then $E_2(L) = (\Delta_2(L))$.

Theorem III b(s) is a unit (respectively, c(t) is a unit) in $\mathbb{Z}[G/G'] \Leftrightarrow$ the longitude of ℓ_2 (respectively, of ℓ_1) lies in G".

Corollary to Theorems I and III I principal $\Leftrightarrow$ the longitudes of L lie in G". "

His only comment on the proof was that a mixture of algebraic and geometric techniques were used. In [73] an argument involving "singular Seifert surfaces" and Alexander duality in S^3 was used to show that their final corollary holds for any homology boundary link. The arguments below, which are of more general applicability, rely instead on equivariant Poincaré duality in the maximal abelian covering, and the Universal Coefficient spectral sequence.

The Main Theorem

If L is a μ-component homology boundary link, so that G maps onto $F(\mu)$, then A(L) maps onto $\Lambda^{\mu} = A(F(\mu))$. In fact it suffices that G maps onto $F(\mu)/F(\mu)''$, the free metabelian group on μ generators, since the

Alexander module depends only on the maximal metabelian quotient of the group. (We shall show below that this condition is also necessary if $\mu = 2$.)

Theorem 1 Let L be a μ-component link of Alexander nullity μ, and let B be the submodule of A(L) generated by the images of the longitudes. Then

(i) B is pseudozero:

(ii) $A(L) \approx tA(L) \oplus \Lambda^{\mu}$ if and only if $E_{\mu}(A(L)/B)$ is principal.

Moreover if $E_{\mu}(A(L)/B)$ is principal then $E_{\mu}(L) = (\Delta_{\mu}(L)).E_0(B)$, B is the maximal pseudozero submodule of A(L), $\mathrm{Ann}(tA(L)/B) = (\Delta_{\mu}(L)/\Delta_{\mu+1}(L))$ and A(L)/B is torsion free as an abelian group.

Proof (i) Since $\alpha(L) = \mu$, every ν-component sublink of L has Alexander nullity ν, and so every 2-component sublink has Alexander polynomial 0. Therefore all the pairwise linking numbers of L are certainly 0, and so the longitudes lie in G'. Let ℓ_i be the image of an ith longitude in G'/G". Since each ith longitude commutes with an ith meridian, $(1-t_i).\ell_i = 0$, and so ℓ_i is in $tG'/G" = tA(L)$. Now $E_{\mu}(L)$ must contain some element δ such that $\varepsilon(\delta) = 1$, since $\varepsilon(E_{\mu}(L)) = \mathbb{Z}$. Let $S = \{\delta^n | n \geq 0\}$. Then $A(L)_S$ is projective by the Corollary of Theorem III.9, so some power of δ annihilates tA(L). Therefore if B is the submodule of A(L) generated by $\ell_1, \ldots, \ell_{\mu}$, Ann B contains $\prod_{i=1}^{\mu}(1-t_i)$ and δ^N for some $N >> 0$. Since $\varepsilon(\delta) = 1$ these elements of Λ have no nontrivial common factor, and so $(\mathrm{Ann}\,B)^{\sim} = \Lambda$.

(ii) Let Y be the closed 3-manifold obtained via (0-framed) surgery on the longitudes of L, and let Y' be its maximal abelian covering space. Then $H_1(Y;\mathbb{Z}) = G/G'$ (since the longitudes are in G') and $H_1(Y;\Lambda) = H_1(Y';\mathbb{Z}) = (G'/G")/B$.

Suppose that $A(L) \approx tA(L) \oplus \Lambda^{\mu}$. Then since $B \subseteq tG'/G'' = tA(L)$, the direct sum splitting of $A(L)$ induces a splitting $H_1(Y;\Lambda) = T \oplus D$, where $T = (tG'/G'')/B$ is a torsion module and D lies in an exact sequence:

$$0 \to D \to \Lambda^{\mu} \to \Lambda \xrightarrow{\varepsilon} \mathbb{Z} \to 0.$$

If $\mu \leq 2$ it follows from Schanuel's lemma that D is stably free, and since it is then of rank at most 1, it must be free. If $\mu \geq 3$, then $e^{\mu-2}D = e^{\mu}\mathbb{Z} = \mathbb{Z}$ and $e^{q}D = 0$ for $q \neq 0, \mu-2$, while there is an exact sequence:

$$0 \to \Lambda \to \Lambda^{\mu} \to e^{o}D \to 0 .$$

By Poincaré duality $H_2(Y;\Lambda) \approx \overline{H^1(Y;\Lambda)} \approx e^{o}D$ and so has projective dimension 1. Therefore the only nonzero entries in the E_2 level of the Universal Coefficient spectral sequence

$$E_2^{pq} = e^{q}H_p(Y;\Lambda) = \mathrm{Ext}^{q}_{\Lambda}(H_p(Y;\Lambda),\Lambda) \Rightarrow H^{p+q}(Y;\Lambda)$$

are $E_2^{o\mu} = e^{\mu}\mathbb{Z} = \mathbb{Z}$, $E_2^{1q} = e^{q}T \oplus e^{q}D$ (for $0 \leq q \leq \mu+1$), $E_2^{20} = e^{o}H_2(Y;\Lambda)$ and $E_2^{21} = e^{1}H_2(Y;\Lambda) = \mathbb{Z}$. It will suffice to prove that $e^{q}T = 0$ for $q \geq 2$, for then $T = tA(L)/B$ will have a square presentation matrix (using Suslin's theorem again) and so $E_{\mu}(A(L)/B) = E_0(T)$ will be principal.

If $\mu \geq 6$ it is clear from the spectral sequence that $e^{q}T = 0$ for $q > 2$. In general, let δ be as in (i) above. Since δ augments to 1, localizing with respect to powers of δ does not affect maps between copies of the augmentation module $\mathbb{Z}$. Since δ annihilates $tA(L)$, and since localization is an exact functor, on localizing the above spectral sequence with respect to powers of δ all the terms $e^{q}T$ are annihilated. Since $H^3(Y;\Lambda) \approx \overline{H_o(Y;\Lambda)} = \mathbb{Z}$, we may conclude that the maps between the copies of $\mathbb{Z}$ corresponding to $e^{\mu}\mathbb{Z}$, $e^{\mu-2}D$ and $e^{1}H_2(Y;\Lambda)$ are what they ought to be. Therefore from the unlocalized spectral sequence it follows that $e^qT = 0$ if $q > 2$, and there is an exact sequence:

$$0 \to e^1T \to H^2(Y;\Lambda) \to e^0H_2(Y;\Lambda) \to e^2T \to 0.$$

Now $H^2(Y;\Lambda) \approx \overline{H_1(Y;\Lambda)} \approx \bar{T} \oplus \bar{D}$, and $e^0H_2(Y;\Lambda) = e^0(\overline{e^0D}) = \overline{e^0e^0}D$. Hence there is an isomorphism $e^1T \approx \bar{T}$ (which is essentially the Blanchfield linking pairing for the cover $Y' \to Y$ [11]) and a short exact sequence

$$0 \longrightarrow D \xrightarrow{\alpha} e^0e^0D \longrightarrow e^2\bar{T} \longrightarrow 0 .$$

On dualizing this sequence it follows that $e^0\alpha : e^0e^0e^0D \to e^0D$ is an isomorphism, since $e^qe^2\bar{T} = 0$ for $q < 2$. Hence $e^0e^0\alpha : e^0e^0D \to e^0e^0e^0e^0D$ is an isomorphism. But e^0e^0D is naturally isomorphic to D (as follows from the exact sequences defining D and presenting e^0D) and hence $e^0e^0\alpha = \alpha$. Therefore $e^2\bar{T} = 0$. (If $\mu \leq 2$, so D is free, this follows more easily, for e^2T is then a pseudozero module with a short free resolution). Hence $e^2T = 0$ and so T has projective dimension at most 1.

It now follows from Lemma III.6 that $E_0(tA(L)) = E_0(tA(L)/B).E_0(B)$, and so $E_\mu(L) = E_0(tA(L)) = (\Delta_\mu(L)).E_0(B)$, since $E_0(tA(L)/B)$ is principal and B is pseudozero.

The argument in the other direction and the remaining assertions are immediate consequences of Theorems III.9, III.10, III.4 Remark 2 and IV.4 (respectively). //

<u>Corollary</u> $E_{\mu-1}(L) = 0$ <u>and</u> $E_\mu(L)$ <u>is principal if and only if</u> $A(L)$ <u>maps onto</u> Λ^μ <u>and the longitudes of</u> L <u>lie in</u> G'', <u>and in this case</u> G'/G'' <u>is torsion free as an abelian group and</u> $\mathrm{Ann}(tG'/G'') = (\Delta_\mu(L)/\Delta_{\mu+1}(L))$. //

<u>Remarks</u> 1. If $E_\mu(L)$ is principal then $\Delta_\mu(L)$ is in $E_\mu(L) = E_0(tA(L))$ which is contained in $\mathrm{Ann}(tA(L))$ by Theorem III.1 and so the arguments

involving δ in (i) above may be simplified.

2. If there is an epimorphism $\eta : G \to F(\mu)/F(\mu)''$ then $\ker \eta = G(\infty)$, since $F(\mu)/F(\mu)''$ is residually nilpotent [145; page 76].

3. The conditions in the corollary on A(L) and on the longitudes of L each imply $E_{\mu-1}(L) = 0$; otherwise all four conditions are independent. For instance the link $L_1 \cup L_2$ of Figure V.1 has $E_1(L_1 \cup L_2) = 0$ and has its longitudes in G'', but $E_2(L) \approx (2-t_1, 1+t_2)(2t_1-1, 1+t_2)$ and $tA(L) = \Lambda_2/(2-t_1, 1+t_2)$ is pseudozero. The link depicted in Figure II.1 is a 2-component homology boundary link for which $E_2(L)$ is not principal.

4. (A late insertion). An unpublished theorem of McIsaac and Webb implies that a metabelian group H such that $H/H' = \mathbb{Z}^\mu$ and $A(H) \approx \Lambda^\mu$ is free metabelian [201]. Hence the above corollary can be restated entirely in terms of the link group.

A Remarkable Example

In this section we give an example of a 2-component ribbon link with trivial components and which has the same Alexander module as a trivial 2-component link, yet which is not even an homology boundary link. Thus it is a counter-example to Smythe's conjecture, although it cannot be distinguished from the trivial 2-component link by the usual metabelian invariants. That it is not an homology boundary link is proven by showing that G/G_ω is one of the nonfree parafree groups of Baumslag [9]. This link has the additional noteworthy properties that its longitudes lie in G'', and that G is a split extension of G/G_ω (and thus is a semi-direct product).

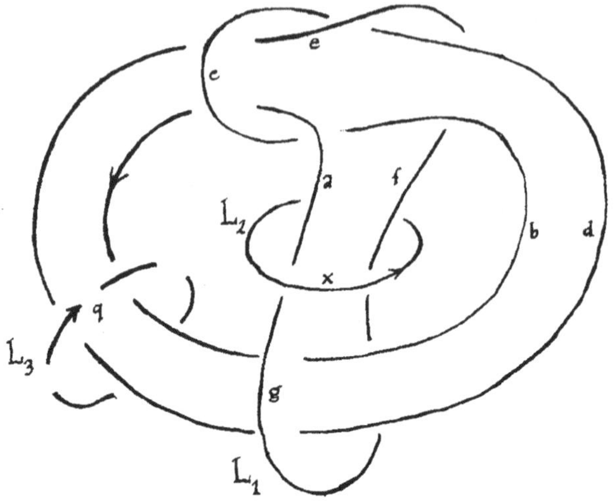

Figure 1

Let $L = L_1 \cup L_2 : 2S^1 \to S^3$ be the link represented by the solid lines in **Figure** 1 and let G be its link group. Then G has a presentation

$$\{ a,b,c,d,e,f,g,x \mid a^{-1}xa.f^{-1}x^{-1}f,\ aba^{-1}.c^{-1},$$

$$b^{-1}x^{-1}fxb.d^{-1}g^{-1}d,\ b^{-1}fb.d^{-1}e^{-1}d,\ cac^{-1}.gb^{-1}g^{-1},$$

$$cec^{-1}.gd^{-1}g^{-1},\ ede^{-1}c^{-1},\ xgx^{-1}a^{-1}\},$$

where a,b,c,d,e,f,g,x represent Wirtinger generators associated with the arcs so labelled in the figure. The longitudes of L are represented by fa^{-1} and by $c^{-1}ga^{-1}eg^{-1}cdb^{-1}xbd^{-1}x^{-1}$. Introduce new generators $\beta,\gamma,\delta,\varepsilon,\phi,\theta$ and new relators $ba^{-1}\beta^{-1}$, $ca^{-1}\gamma^{-1}$, $da^{-1}\delta^{-1}$, $ea^{-1}\varepsilon^{-1}$, $fa^{-1}\phi^{-1}$, $ga^{-1}\theta^{-1}$. Then the above presentation of G is Tietze-equivalent to

$$\{ a,x,\beta,\gamma,\delta,\varepsilon,\phi,\theta \mid x\phi x^{-1}\phi^{-1},\ a\beta a^{-1}\gamma^{-1},\ \beta^{-1}x^{-1}\phi ax\beta\delta^{-1}a^{-1}\theta^{-1}\delta,$$

$$\beta^{-1}\phi a\beta\delta^{-1}a^{-1}\varepsilon^{-1}\delta,\ \gamma a\gamma^{-1}\theta\beta^{-1}a^{-1}\theta^{-1},\ \gamma a\varepsilon\gamma^{-1}\theta\delta^{-1}a^{-1}\theta^{-1},$$

$$\varepsilon a\delta\varepsilon^{-1}a^{-1}\gamma^{-1},\ x\theta ax^{-1}a^{-1} \}$$

and the longitudes are represented by ϕ and $a^{-1}\gamma^{-1}\theta\varepsilon\theta^{-1}\gamma a\delta\beta^{-1}x\beta\delta^{-1}x^{-1}$.

Clearly $\beta,\gamma,\delta,\varepsilon,\phi,\theta$ represent elements of G', and $\gamma\varepsilon\gamma^{-1}\theta\delta^{-1}\beta\theta^{-1}$ represents the trivial element of G, so $\beta\varepsilon\delta^{-1}$ represents an element of G". Thus G/G" has a presentation

$$\{a,x,\beta,\gamma,\delta,\varepsilon,\phi,\theta \mid x\phi x^{-1}\phi^{-1},\ a\beta a^{-1}\gamma^{-1},\ x^{-1}\phi a x\varepsilon^{-1}a^{-1}\theta^{-1}\varepsilon,$$
$$\phi a\varepsilon^{-1}a^{-1},\ \gamma a\gamma^{-1}\beta^{-1}\theta a^{-1}\theta^{-1},\ \beta\varepsilon\delta^{-1},\ \varepsilon a\delta\varepsilon^{-1}a^{-1}\gamma^{-1},$$
$$\theta a x^{-1}a^{-1}x,\ [[\ ,\ ],[\ ,\ ]]\}$$

where $[[\ ,\ ],[\ ,\ ]]$ denotes the set of all commutators of commutators in the generators. It follows that $\phi\theta a.a^{-1}\phi^{-1}\theta^{-1}\varepsilon$ represents 1 in G/G", and so ε and hence ϕ and $\delta\beta^{-1}$ represent 1 in G/G". (Consequently the longitudes of L lie in G"). The presentation for G/G" is therefore equivalent to

$$\{a,x,\beta,\gamma,\theta \mid a\beta a^{-1}\gamma^{-1},\ \gamma a\gamma^{-1}\beta^{-1}\theta^{-1}a^{-1}\theta^{-1},\ \theta a x^{-1}a^{-1}x,\ [[\ ,\ ],[\ ,\ ]]\}.$$

Then $a\beta\gamma^{-1}\beta^{-1}\theta a^{-1}\theta^{-1}$ represents 1 in G", and so $\gamma\theta a\theta^{-1}a^{-1}$ and $\beta a^{-1}\theta a\theta^{-1}$ do also. Thus this presentation is equivalent to $\{a,x,\theta \mid \theta a x^{-1}a^{-1}x,\ [[\ ,\ ],[\ ,\ ]]\}$ and finally to $\{a,x \mid [[\ ,\ ],[\ ,\ ]]\}$. Thus $G/G" \approx F(2)/F(2)"$, the free metabelian group on 2 generators.

As L is visibly a ribbon link, the projection of G onto G/G_ω factors through H(R), where H(R) is the group with presentation $\{a,b,c,g,x \mid aba^{-1}c^{-1}, cac^{-1}.gbg^{-1}, xgx^{-1}.a^{-1}\}$ (so $H(R) = G/\langle\langle \beta\delta^{-1}, \varepsilon, \phi\rangle\rangle$). This presentation is equivalent to $\{a,b,x,t \mid abab^{-1}a^{-1}, x^{-1}axb^{-1}x^{-1}a^{-1}x, t.x^{-1}a^{-1}xaba^{-1}\}$ and thus to $\{a,x,t \mid t.x^{-1}a^{-1}xa.t^{-1}ata^{-1}\}$. Thus H(R) is isomorphic to the group $G_{-1,1}$ of Baumslag, which is parafree (so $H(R)_\omega = 1$) but not free [9]. Thus L cannot be an homology boundary link (for otherwise an epimorphism $G \to F(2)$ would induce an isomorphism $G/G_\omega \approx F(2)$ and H(R) would then be free). Since the longitudes of L are in G" and since the meridians a,x map to a generating set for $F(2)/F(2)''$ this link is a counter-example to Questions 1 and 2 of [174]. (The three component link $\hat{L} = L \cup L_3$ is an homology boundary link whose longitudes lie in G''. Since L is not a boundary link $\hat{L}$ cannot be a boundary link. Thus even if the link of question 1 of [174] is assumed to be a homology boundary link, it need not be a boundary link. Lambert had earlier constructed an example of this kind, with only 2 components, but 51 crossings! [110].) The link is also a counter-example to lemma 8 of [63], which implies that a slice link with unknotted components such that G/G'' is generated by 2 meridians is a boundary link.

From the presentation first given for G and H(R) it follows easily that there is a splitting homomorphism $H(R) \to G$ for the projection, so G is a semidirect product $G \approx G_\omega \rtimes H(R)$. (It may be shown that the group of the link of Figure V.1 is not such a semidirect product [76].) Consequently $G_\omega = [G, G_\omega]$. It seems unlikely that G_ω is perfect (equal to its own commutator subgroup), although as G_ω is the normal closure of the single element represented by ε, G_ω/G_ω' is a cyclic $\mathbb{Z}[G/G_\omega]$-module. Is the link L interchangeable, that is, is there a homeomorphism $h : S^3 \to S^3$ such that $h \circ L_1 = L_2$ and $h \circ L_2 = L_1$?

This example was motivated by McMillan's example of an embedding $W : S^1 \vee S^1 \to S$ such that $\pi_1(S^3 - \operatorname{im} W) \approx H(R)$ [127], and was constructed by finding a Wirtinger presentation of deficiency 2 for H(R) and then forming a ribbon link whose ribbon group had that presentation.

2-Component Links

For links with 2 components, the above results can be strengthened in various ways.

Theorem 2 *Let L be a 2-component link. Then the following are equivalent*

(i) G *maps onto* $F(2)/F(2)''$;

(ii) A(L) *maps onto* Λ_2^2;

(iii) G'/G'' *maps onto* Λ_2.

Proof (i) $\Longrightarrow$ (ii) Crowell showed in [37] that A(L) depends functorially on G/G''. An epimorphism : $G \to F(2)/F(2)''$ induces a map : $G/G'' \to F(2)/F(2)''$ which splits since $F(2)/F(2)''$ is free metabelian. Hence Λ_2^2, the Alexander module of a (trivial) 2-component link with group F(2), is a direct summand of A(L).

(ii) $\Longrightarrow$ (iii) The Crowell exact sequence gives rise to an exact sequence

$$0 \to (G'/G'')/(tG'/G'') \to A(L)/tA(L) \to \Lambda \to \mathbb{Z} \to 0$$

and so $(G'/G'')/(tG'/G'')$ is stably free, by Schanuel's lemma, and hence free, since it is of rank 1.

(iii) $\Longrightarrow$ (i) Let $H = (G/G'')/(tG'/G'')$. Then $H'/H'' \approx \Lambda_2$. The extensions of $\mathbb{Z}^2$ by Λ_2 are classified by $H^2(\mathbb{Z}^2;\Lambda_2) \approx \mathbb{Z}$, and it is easily checked that if

$$1 \to \Lambda_2 \to Ex_n \to \mathbb{Z}^2 \to 1$$

is an extension corresponding to $n \in \mathbb{Z}$, then $Ex_n/Ex_n' \approx \mathbb{Z}^2 \oplus (\Lambda_2/n\Lambda_2)$ and that $Ex_{-1} \approx Ex_1$. It follows that $H \approx F(2)/F(2)''$ and hence that G maps onto $F(2)/F(2)''$. //

Remark For $\mu > 2$, the only difficulty is deducing from "A(L) maps onto Λ_μ^μ" that "G maps onto $F(\mu)/F(\mu)''$ " is the lack of an adequate cancellation theorem for nonprojective modules (Schanuel's lemma shows that $((G'/G'')/(tG'/G'')) \oplus \Lambda_\mu^\mu \approx (F(\mu)'/F(\mu)'') \oplus \Lambda_\mu^\mu$, and the classification of extensions of $\mathbb{Z}^\mu$ by $F(\mu)'/F(\mu)''$ is easy). See Remark 4 on page 71.

Theorem 3 Let L be a 2-component link with Alexander nullity 2. Then the following are equivalent:

(i) $E_2(L)$ is principal ;

(ii) $E_1(G'/G'')$ is principal ;

(iii) $p.d.A(L) \leqslant 1$;

(iv) $p.d.G'/G'' \leqslant 1$.

Proof (i) $\Longrightarrow$ (ii),(iii),(iv). If $E_2(L)$ is principal then $A(L) \approx tA(L) \oplus \Lambda_2^2$ and $tA(L)$ has a square presentation matrix by Theorem III.9, whence (iii) holds, and $G'/G'' \approx tA(L) \oplus \Lambda_2$ by Theorem 2 so (ii) and (iv) hold.

(ii) $\Longrightarrow$ (i),(iii),(iv). This is similar.

(iii) $\Longleftrightarrow$ (iv). This follows from Lemma IV.2.

(iv) $\Longrightarrow$ (ii). Suppose that $p.d.G'/G'' \leqslant 1$. Then the longitudes of L must lie in G'', by Theorem III.10 and Theorem 1. Let Y be the closed 3-manifold obtained by surgery on the longitudes of L. Then $H_1(Y;\Lambda) \approx G'/G''$. It follows as in Theorem 1 that there is an exact sequence

$$H^2(Y;\Lambda) \to e^0H_2(Y;\Lambda) \to e^2(G'/G'')$$

and that $H^2(Y;\Lambda) \approx \overline{H_1(Y;\Lambda)}$ by Poincaré duality. Since $H_2(Y;\Lambda)$ is of rank 1, $e^0H_2(Y;\Lambda) \approx \Lambda$. Therefore since $p.d.G'/G'' \leqslant 1$, $G'/G'' = \overline{H^2(Y;\Lambda)}$ maps onto $\overline{\Lambda} = \Lambda$, so $G'/G'' \approx \Lambda \oplus tG'/G''$, $p.d.tG'/G'' \leqslant 1$, and (ii) follows from Theorem III.9. //

Our next result includes the assertion of Crowell and Brown on the structure of $(E_2(L):(\Delta_2(L))$ for L a 2-component homology boundary link.

Theorem 4 Let L be a 2-component link with Alexander nullity 2, and let B be the submodule of A(L) generated by the longitudes. Then there

<u>are elements</u> $b_1(t_2)$, $b_2(t_1) \in \Lambda_2$ <u>with</u> $b_1(1) = b_2(1) = 1$ <u>such that</u> $B \approx (\Lambda_2/(b_1(t_2), t_1 - 1)) \oplus (\Lambda_2/(b_2(t_1), t_2 - 1))$. <u>Hence</u> $\text{Ann } B = E_0(B) = (b_1(t_2) + b_2(t_1) - 1, (t_1 - 1)b_2(t_1), (t_2 - 1)b_1(t_2))$, <u>and</u> B <u>is torsion free as an abelian group.</u>

<u>Proof</u> Let B_1 and B_2 be the cyclic submodules of B generated by the first and second longitudes of L respectively. Since Λ_2 is a regular domain of dimension 3, and since $\text{p.d.}G'/G'' \leq 2$, G'/G'' contains no submodules of finite length. Therefore the same is true of B_1 and B_2, so they each have projective dimension less than 3, by Remark 2 after Theorem III.11 Therefore in particular B_2 has a finite free resolution

$$0 \to \Lambda_2^b \to \Lambda_2^{b+1} \xrightarrow{Q} \Lambda_2 \to B_2 \to 0.$$

There is also a short exact sequence:

$$0 \to \Lambda_2 \xrightarrow{t_2 - 1} \Lambda_2 \xrightarrow{\eta} \Lambda_1 \to 0 .$$

Since $t_2 - 1$ annihilates B_2, it follows that $\text{Tor}_1^{\Lambda_2}(\Lambda_1, B_2) \approx \Lambda_1 \otimes_{\Lambda_2} B_2 = B_2$. Then there are exact sequences:

$$0 \to \ker \tilde{Q} \to \Lambda_1^{b+1} \to \Lambda_1^1 \to B_2 \to 0$$

and

$$0 \to \Lambda_1^b \to \ker \tilde{Q} \to B_2 \to 0,$$

where $\tilde{Q}$ is the reduction of Q under the ring homomorphism η. By Schanuel's lemma and Suslin's theorem $\ker \tilde{Q}$ is a free Λ_1-module, of rank b, and so B_2, considered as a Λ_1-module, has a square presentation matrix. Therefore the annihilator ideal of B_2 in Λ_1 is principal, generated by some element $b_2(t_1)$. Since $\Delta_1(L) = 0$ the longitudes of L are in G', that is are null-homologous in X, and so $b_2(t_1)$ must augment to a generator ± 1 of $\mathbb{Z}$, which may be assumed to be $+1$. Therefore B_2 is torsion free as an abelian group, by Theorem IV.4. Considering B_2 now as a Λ_2-module, we conclude that

Ann $B_2 = (b_2(t_1), t_2 - 1)$ where $b_2(1) = 1$. Similarly Ann $B_1 = (b_1(t_2), t_1 - 1)$ for some $b_1(t_2)$ such that $b_1(1) = 1$, and B_1 is torsion free as an abelian group.

Let ℓ_1 and ℓ_2 be generators of B_1 and B_2 respectively. Suppose that $a_1(t_1,t_2), a_2(t_1,t_2) \in \Lambda_2$ are such that $a_1(t_1,t_2).\ell_1 + a_2(t_1,t_2).\ell_2 = 0$. Then $a_1(t_1,t_2).\ell_1 = a_1(t_1,t_2).b_2(t_1).\ell_1$ (since $b_2(t_1) \equiv 1$ modulo $(t_1 - 1)) = 0$ (since $b_2(t_1).\ell_2 = 0$). Therefore $a_1(t_1,t_2)$ is Ann B_1, and similarly $a_2(t_1,t_2)$ is in Ann B_2, and so $B = B_1 \oplus B_2 \approx (\Lambda_2/(b_1(t_2), t_1 - 1) \oplus (\Lambda_2/(b_2(t_1), t_2 - 1))$. In particular, B is torsion free as an abelian group.

Let $p = b_1(t_2) + b_2(t_1) - 1$, $q = (t_1 - 1)b_2(t_1)$, $r = (t_2 - 1)b_1(t_2)$, $s = (t_1 - 1)(t_2 - 1)$, $t = b_1(t_2)b_2(t_1)$, $b_1' = (b_1(t_2) - 1/t_2 - 1)$ and $b_2' = (b_2(t_1) - 1/t_1 - 1)$. Then $E_0(B) = (q,r,s,t) = (p,q,r)$ since $p = t - s.b_1'.b_2'$ and $t = b_2(t_2).p - b_2'.q$ and $s = -s.p + (t_2 - 1).q + (t_1 - 1).r$. Clearly also $E_0(B) \subseteq \text{Ann } B = \text{Ann } B_1 \cap \text{Ann } B_2$.

Suppose that $a(t_1,t_2)$ is in Ann B. Then

$$\begin{aligned} a(t_1,t_2) &= m(t_1,t_2).(t_1 - 1) + n(t_1,t_2).b_1(t_2) \qquad \text{(since it is in Ann } B_1\text{)} \\ &= m(t_1,1).(t_1 - 1) + n(t_1,1).(1 - b_2(t_1)) + (m(t_1,t_2) \\ &\quad - m(t_1,1)).(t_1 - 1).b_2(t_1) + (t_1 - 1).(m(t_1,t_2) - m(t_1,1)).b_1(t_2) \\ &\quad - (m(t_1,t_2) - m(t_1,1)).p + (n(t_1,t_2) - n(t_1,1)).b_1(t_2) + n(t_1,1).p. \end{aligned}$$

Therefore (invoking the Remainder Theorem to conclude $t_2 - 1$ divides $m(t_1,t_2) - m(t_1,1)$ and $n(t_1,t_2) - n(t_1,1)$) it follows that $m(t_1,1)(t_1 - 1) + n(t_1,1).(1 - b_2(t_1))$ is also in Ann B, and so equals some $u(t_1,t_2).(t_2 - 1) + v(t_1,t_2).b_2(t_1)$ (since it is in Ann B_2). On setting $t_2 = 1$ it follows that $m(t_1,1).(t_1 - 1) + n(t_1,1).(1 - b_2(t_1)) = v(t_1,1).b_2(t_1)$, and on setting $t_1 = 1$, it follows that $v(1,1) = 0$, so $v(t_1,1) = w(t_1).(t_1 - 1)$ (by the Remainder Theorem again). Thus

$$a(t_1,t_2) = (m(t_1,t_2) - m(t_1,1) + w(t_1)).(t_1 - 1).b_2(t_1)$$
$$+ (t_1 - 1).(m(t_1,t_2) - m(t_1,1).b_1(t_2) + (n(t_1,t_2)$$
$$- n(t_1,1)).b_1(t_2) + (n(t_1,1) + m(t_1,1) - m(t_1,t_2)).p.$$

Therefore $a(t_1,t_2)$ is in (p,q,r) and so Ann B $= E_0(B) = (p,q,r)$

$$= (b_1(t_2) + b_2(t_1) - 1, \quad (t_1 - 1).b_2(t_1), \quad (t_2 - 1).b_2(t_2)). \ //$$

Corollary *If* G *maps onto* F(2)/F(2)" *then* G'/G" *is torsion free as an abelian group.*

Proof For then by Theorem IV.4 and Theorem 1 (G'/G")/B is torsion free as an abelian group. //

Meridians

We conclude this chapter with a comment on 2-component boundary links. To show that an homology boundary link is not a boundary link, even though the μth Alexander ideal is principal, it must be shown that no set of μ meridians maps to a set of generators for $G/G_\omega = F(\mu)$. There is an algorithm due to Whitehead for deciding whether a given set of elements of $F(\mu)$ generates the group [123; page 166], but here the possibility of replacing elements by conjugates must also be allowed. If $\mu = 2$, a theorem of Nielsen leads to a simple answer.

Definition The element w_1 in $F(\mu)$ is *primitive* if there are elements $w_2,\ldots,w_\mu$ in $F(\mu)$ such that $\{w_1,\ldots,w_\mu\}$ generates $F(\mu)$; equivalently, if there is an automorphism ψ of $F(\mu)$ such that $\psi(w_1) = x_1$, where $F(\mu)$ is the free group on the letters $\{x_1,\ldots,x_\mu\}$.

Theorem (Nielsen [123; page 169]). *There is at most one conjugacy class of primitive elements of* F(2) *with given image in* $F(2)/F(2)' = \mathbb{Z}^2$.

Therefore if w_1 and w_2 in F(2) generate F(2) modulo F(2)', some conjugates of w_1 and w_2 generate F(2) if and only if they are each primitive. For clearly this is necessary. Suppose w_1 and w_2 are each primitive. Then after an automorphism ψ of F(2) it may be assumed that $\psi(w_1) = x_1$ and $\psi(w_2) = x_1^a x_2^b$ modulo F(2)'. Since w_1 and w_2 generate modulo F(2)', b must be ± 1. But the element $x_1^a x_2^{\pm 1}$ is clearly primitive, and so $\psi(w_2) = z x_1^a x_2^{\pm 1} z^{-1}$ for some z. Therefore w_1 and $(\psi^{-1}(z))^{-1} w_2 (\psi^{-1}(z))$ generate F(2).

We shall illustrate this result by showing that the link of Figure II.1 is not a boundary link. (We thereby avoid appealing to a Seifert surface argument, to show that the second Alexander ideal of a 2-component boundary link is principal). The ribbon group of this link has a presentation

$$\{a,w,x,y,z \mid axa^{-1} = y,\ wyw^{-1} = z,\ zwz^{-1} = x\}$$

which is Tietze-equivalent to

$$\{a,w,z \mid wazwz^{-1}a^{-1}w^{-1} = z\}$$

and hence to $\{b,w \mid \phi\}$ where $b = waz$. Thus the ribbon group is free (so the link is an homology boundary link) and the images of the meridians a and w are represented by the words $w^{-1}b^2w^{-1}b^{-1}$ and w in the free generators b and w. Since a and bw^{-2} have the same image in G/G', and since bw^{-2} is clearly primitive, the link can only be a boundary link if a is conjugate to bw^{-2}. But the words $w^{-1}b^2w^{-1}b^{-1}$ and bw^{-2} are clearly each cyclically reduced, and of distinct lengths, and so do not represent conjugate elements of the free group $\{b,w \mid \phi\}$. [123; page 36].

Bachmuth has shown that the analogue of Nielsen's theorem holds also for the free metabelian group of rank 2, F(2)/F(2)" [6]. His results have been used by Brown to prove that the Λ_μ-module $F(\mu)'/F(\mu)''$ is not the direct sum of two proper submodules [17].

Osborne and Zieschang have given a simple procedure for finding a primitive word in the coset of $x_1^m x_2^n$ modulo $F(2)'$ whenever $(m, n) = 1$ [208]. Their formulae apply also in the metabelian case.

Suppose finally that L is a 2-component link such that G maps onto $F(2)/F(2)''$. If L is a boundary link then there is a pair of meridians in G which maps to a generating set for $F(2)/F(2)''$. In general when is this the case? Is it so if $E_2(L)$ is principal, or conversely?

(Remark. We should have observed earlier that for a μ-component boundary 1-link L, the ideal $E_\mu(L)$ is principal. This follows from the Corollary to Theorem 1 and the fact that the longitudes lie in $G'_\omega \subseteq G''$ (page 14) or more directly from the Mayer-Vietoris sequence of the maximal abelian cover determined by a set of disjoint Seifert surfaces, which gives a square presentation matrix for tG'/G''.)

CHAPTER VII SUBLINKS

In this chapter we shall relate the Alexander invariants of a link to those of its sublinks. For the first Alexander polynomial this was done by Torres, who used properties of Wirtinger presentations of link groups to establish two conditions on Δ_1 [189]. Sato showed that one could derive Torres' second condition from the Wang sequence and excision [165]. Traldi has extended the second condition to the higher Alexander ideals [190]. We shall show that Sato's argument applies equally well in this case. (Torres' first condition can be deduced from the second condition and duality if all the linking numbers are nonzero, and then the general case follows by a simple argument due to Fox and Torres [52]). We shall give some simple consequences of the Torres conditions and state without proof much stronger results recently announced by Traldi [191, 192].

The Torres conditions (for $\mu = 1$) serve to characterize the first Alexander polynomial of a knot. (This was done much earlier by Seifert [167]). Bailey and Levine have shown that they characterize the first Alexander polynomial of a 2-component link with linking number 0 and ± 1 respectively, while Kidwell has shown that for linking number 3 and under restrictions on the "order" of the link further conditions are necessary [7, 99, 117]. We shall show that without any such restrictions the Torres conditions are in general not sufficient in the 2-component case. Our theorem invokes a derivative of the Alexander polynomial that Murasugi had earlier shown was an invariant of certain link homotopies, and so we shall sketch a proof of Murasugi's result.

The Conditions of Torres and Traldi

Let L be a μ-component link and let ℓ_{ij} be the linking number of the ith and j^{th} components. (Recall $\ell_{ii} = 0$ for all i). Let $\phi: \Lambda_\mu \to \Lambda_{\mu-1}$ be the homomorphism sending t_i to t_i for $i < \mu$ and sending t_μ to 1. Then the two conditions of Torres may be stated as follows:

(1) If $\mu = 1$, $\overline{\Delta_1(L)} = t^{2a}\Delta_1(L)$ for some a ;

if $\mu > 1$, $\overline{\Delta_1(L)} = (-1)^\mu (\prod_{1\leq i\leq\mu} t_i^{b_i})\Delta_1(L)$ where

$$b_i \equiv 1 - \sum_{1\leq j\leq\mu} \ell_{ij} \text{ modulo } (2).$$

(2) If $\mu = 1$, $\phi(E_1(L)) = \mathbb{Z}$;

if $\mu > 1$, $\phi(E_1(L)) = (\prod_{1\leq i\leq\mu} t_i^{\ell_{i\mu}} - 1)\, E_1(L_{\hat{\mu}})$

where $L_{\hat{\mu}}$ is the sublink obtained by deleting the μth component of L.

(Note that the first condition does not depend on the choice of first Alexander polynomial, and that deleting other components of L leads to conditions similar to (2)).

The first condition may be restated in the following slightly weaker form:

(1)' The principal ideals $(\overline{\Delta_1(L)})$ and $(\Delta_1(L))$ are equal.

We shall prove the following extension of (1)', first obtained by Blanchfield. (Our argument is related to his).

<u>Theorem 1</u> (Blanchfield [11]) <u>For each</u> $i \geq 1$, <u>the principal ideals</u> $(\overline{\Delta_i(L)})$ <u>and</u> $(\Delta_i(L))$ <u>are equal</u>.

<u>Proof</u> Since Λ is a factorial domain it will suffice to show that $\overline{\Delta_i(L)}$ and $\Delta_i(L)$ have the same irreducible factors. Let $\mathfrak{p} = (p)$ be a height 1 prime ideal, generated by an irreducible element p. If $\bar{\mathfrak{p}} = \mathfrak{p}$, then p^a divides $\overline{\Delta_i(L)}$ if and only if it divides $\Delta_i(L)$. So we may assume that $\bar{\mathfrak{p}} \neq \mathfrak{p}$, and hence that $t_i - 1$ is a unit in the localization $\Lambda_{\mathfrak{p}}$. On localizing the long exact sequence of equivariant homology for the maximal abelian cover of the pair $(X, \partial X)$, and on observing that $\prod_{1 \leq i \leq \mu} (t_i - 1)$ annihilates $H_*(\partial X : \Lambda)$, we conclude that $M_{k\mathfrak{p}} = H_k(X;\Lambda)_{\mathfrak{p}}$ is isomorphic to $H_k(X, \partial X;\Lambda)_{\mathfrak{p}}$ for all k. By Poincaré duality $\overline{H_1(X,\partial X;\Lambda)_{\bar{\mathfrak{p}}}}$ is isomorphic to $H^2(X;\Lambda)_{\mathfrak{p}}$. The Universal Coefficient spectral sequence then gives an exact sequence

$$0 \to e^1 M_1 \to \overline{M_{1\bar{\mathfrak{p}}}} \to M_{2\mathfrak{p}} \to 0$$

Since $M_{2\mathfrak{p}}$ is torsion free and rank $\overline{M_{1\bar{\mathfrak{p}}}}$ = rank $M_{1\mathfrak{p}}$ = rank $M_{2\mathfrak{p}}$, there is an isomorphism $e^1(tM_{1\mathfrak{p}}) = e^1 M_{1\mathfrak{p}} \approx \overline{tM_{1\bar{\mathfrak{p}}}}$. If N is a finitely generated torsion module over a principal ideal domain, there is an unnatural isomorphism $N \approx e^1 N$, by the structure theorem for such modules. The theorem follows. //

We shall use localization in conjunction with duality again in Chapter IX in order to construct an invariant of link concordance.

For knots it is not necessary to localize, as $H_1(X;\Lambda) = H_1(X,\partial X;\Lambda)$ and so the Universal Coefficient spectral sequence and duality imply directly that there is an isomorphism $e^1 H_1(X;\Lambda) \approx \overline{H_1(X;\Lambda)}$. Since $H_1(X;\Lambda)$ has a short free resolution with a square presentation matrix, it follows

$\overline{E_i(L)} = E_i(L)$ for all i, if $\mu = 1$. This is also true for $\mu > 1$, if i=1, as condition (1)' is equivalent to " $\overline{E_1(L)} = E_1(L)$ " by part (i) of Theorem IV.3. Is it true in general? The Steinitz-Fox-Smythe invariants may be used to show that there are knots for which $H_1(X;\Lambda)$ is not isomorphic to $\overline{H_1(X;\Lambda)}$ and hence which are noninvertible [51, 92].

If $\mu = 1$ or if $\mu \geq 2$ and all the linking numbers ℓ_{ij} are nonzero, conditions (1)' and (2) together imply (1). For (1)' is equivalent to $\overline{\Delta_1(L)} = u.\Delta_1(L)$ for some unit $u = (-1)^s \prod_{1 \leq i \leq \mu} t_i^{b_i}$ in Λ_μ. If $\mu = 1$, (2) implies that $\varepsilon(u) = 1$ so that $u = t^b$. Since $\Delta_1(L)\ (1) = \pm 1$ implies that $\Delta_1(L)(-1)$ is odd (and hence nonzero), it then follows that $\Delta_1(L)(-1) = \overline{\Delta_1(L)(-1)} = (-1)^b\ \Delta_1(L)(-1)$, so that $b = 2a$ for some a. In general $\bar{\Delta} = u\Delta$ implies that $\phi(\bar{\Delta}) = \phi(u)\phi(\Delta)$, so if $\mu = 2$ and $\ell = \ell_{12}$

$$(t_1^{\ell} - 1)\ \overline{\Delta_1(L_1)} = t_1^{\ell - 1}\ \phi(u)(t_1^{\ell} - 1)\ \Delta_1(L_1)$$

while if $\mu > 2$ $(\Pi - 1)\overline{\Delta_1(L_\mu^{\wedge})} = -\Pi\phi(u)(\Pi - 1)\Delta_1(L_\mu^{\wedge})$ where $\Pi = \prod_{1 \leq i \leq \mu} t_i^{\ell_{i\mu}}$.

If all the linking numbers ℓ_{1i} are nonzero, a simple induction now gives (1). Otherwise, adjoin a new component K_o such that $\ell_{oi} = \ell(K_o, L_i)$ is nonzero for $1 \leq i \leq \mu$, and let L^+ be the $(\mu+1)$-component link $K_o \amalg L$. We may now use the above argument to conclude $\Delta_1(L^+)(t_o, \ldots t_\mu)$ satisfies (1), so that $\overline{\Delta_1(L^+)} = (-1)^{\mu+1}\ (\prod_{o \leq i \leq \mu} t_i^{c_i})\Delta_1(L^+)$ with

$$c_i \equiv 1 - \sum_{o \leq j \leq \mu} \ell_{ij} \quad \text{modulo } (2).$$

On applying (2) to the link L obtained by deleting the component K_o of L^+, we see that

$$\overline{\left(\prod_{1\leq i\leq \mu} t_i^{\ell_{io}} - 1\right)}\ \Delta_1(L) = (-1)^{\mu+1}\left(\prod_{1\leq i\leq\mu} t_i^{c_i}\right)\left(\prod_{1\leq i\leq\mu} t_i^{\ell_{io}} - 1\right)\Delta_1(L)$$

and hence, on cancelling the nonzero factor $\left(\prod_{1\leq i\leq\mu} t_i^{\ell_{io}} - 1\right)$ from both sides, $-\overline{\Delta_1(L)} = (-1)^{\mu+1} \prod_{1\leq i\leq\mu} t_i^{b_i}\Delta_1(L)$, where $b_i = c_i + \ell_{io} \equiv 1 - \sum_{1\leq j\leq\mu} \ell_{ij}$ modulo (2). Thus (1) is proven. (This derivation of (1) from (2) and duality is due to Fox and Torres [52]. In [189], Torres had shown that $\overline{\Delta_1(L)} = (-1)^{\mu} \prod_{1\leq i\leq\mu} t_i^{n_i}\Delta_1(L)$ for some integers n_i, but he did not determine the values of n_i modulo (2) there).

Using methods similar to those of Torres, Traldi extended condition (2) to the higher Alexander ideals. We shall state his results in the next theorem, but we shall use "pure algebraic topology" instead, following Sato's proof of (2) [165]. For the remainder of this section we shall assume that $\mu > 1$, as (2) is clear in the case of a knot. (In Chapter IV we showed that $\varepsilon(E_\mu(L)) = \mathbb{Z}$ for any μ-component link L. The argument of the following theorem may be applied also in the knot case, with slight changes).

<u>Theorem</u> 2 (Traldi [190]) <u>Let</u> L <u>be a</u> μ-<u>component link, with</u> $\mu > 1$. <u>Then</u>

(i) $\phi(E_1(L)) = (\Pi - 1)\ E_1(L_{\hat\mu})$;

(ii) $E_{k-1}(L_{\hat\mu}) + (\Pi - 1)\ E_k(L_{\hat\mu}) \subsetneq \phi(E_k(L)) \subsetneq E_{k-1}(L_{\hat\mu}) + I_{\mu-1}E_k(L_{\hat\mu})$

<u>for each</u> $k \geqslant 2$.

(Here as before $\Pi = \prod_{1\leq i\leq\mu} t_i^{\ell_{i\mu}}$ and $L_{\hat\mu}$ is the sublink obtained by deleting the last component of L).

Proof Let X be the exterior of L and Y the exterior of $L_{\hat{\mu}}$, and choose a basepoint $*$ in X. Let $\tilde{X}$ be the covering space of X induced by the maximal abelian cover $P_Y : Y' \to Y$ of Y. We shall let $H_*(X;\Lambda_{\mu-1})$ denote the equivariant homology of $\tilde{X}$, and similarly for the pairs $(\tilde{X}, P_Y^{-1}(*))$ and $(Y', \tilde{X})$.

The cover $r:(X',p^{-1}(*)) \to (\tilde{X},P_Y^{-1}(*))$ is infinite cyclic, and so is induced by a map from $\tilde{X}$ to S^1. The Cartan-Leray spectral sequence for r is just the Wang sequence for the fibration $X' \to \tilde{X} \to S^1$, and gives an exact sequence

$$A(L) \xrightarrow{t_\mu - 1} A(L) \longrightarrow H_1(X,*;\Lambda_{\mu-1}) \longrightarrow 0$$

Hence we shall abbreviate $H_1(X,*;\Lambda_{\mu-1})$ as $\phi A(L)$. The long exact sequence of equivariant homology for the triple $(Y',\tilde{X},P_Y^{-1}(*))$ gives an exact sequence.

$$H_2(Y;\Lambda_{\mu-1}) \longrightarrow H_2(Y,X;\Lambda_{\mu-1}) \xrightarrow{\partial} \phi A(L) \longrightarrow A(L_{\hat{\mu}}) \longrightarrow H_1(Y,X;\Lambda_{\mu-1}).$$

By excision $H_1(Y,X;\Lambda_{\mu-1}) = 0$ and $H_2(Y,X;\Lambda_{\mu-1})$ is isomorphic to $\Lambda_{\mu-1}/(\Pi-1)$.

Suppose first that $H_2(Y;\Lambda_{\mu-1}) = 0$, in other words that $E_1(L_{\hat{\mu}}) \neq 0$. Then $A(L_{\hat{\mu}})$ has a short free resolution, and there is a short exact sequence

$$0 \longrightarrow \Lambda_{\mu-1}/(\Pi-1) \longrightarrow \phi A(L) \longrightarrow A(L_{\hat{\mu}}) \longrightarrow 0.$$

Since $\phi(E_1(L)) = E_1(\phi A(L))$, (i) now follows from Lemma III.6. If $H_2(Y;\Lambda_{\mu-1}) \neq 0$ (i) is trivially true, as both sides of the equation are then 0.

In general there is an exact sequence

$$0 \longrightarrow \Lambda_{\mu-1}/J \longrightarrow \phi A(L) \longrightarrow A(L_{\hat{\mu}}) \longrightarrow 0$$

where J is an ideal containing $\Pi-1$. Hence if P is a presentation matrix for $A(L_{\hat{\Omega}})$, $\phi A(L)$ has a presentation matrix of the form $\begin{pmatrix} P & K^* \\ 0 & J^* \end{pmatrix}$ where K^* and J^* are column vectors and the entries of J^* generate the ideal J. Hence $\phi(E_k(L)) = E_k(\phi A(L))$ contains $E_0(\Lambda_{\mu-1}/J)E_k(L_{\hat{\Omega}})$ and $E_1(\Lambda_{\mu-1}/J)E_{k-1}(L_{\hat{\Omega}})$ and so $E_{k-1}(L_{\hat{\Omega}}) + (\Pi-1)E_k(L_{\hat{\Omega}}) \subseteq \phi(E_k(L))$. Similarly $\phi(E_k(L)) \subseteq E_{k-1}(L_{\hat{\Omega}}) + (K,J)E_k(L_{\hat{\Omega}})$. Since $\mathbb{Z} \otimes \phi A(L) = \mathbb{Z}^{\mu}$ and $\mathbb{Z} \otimes A(L_{\hat{\Omega}}) = \mathbb{Z}^{\mu-1}$, the column $\begin{pmatrix} K^* \\ J^* \end{pmatrix}$ is in the span of the columns of $\begin{pmatrix} P \\ 0 \end{pmatrix}$, modulo $I_{\mu-1}$. Hence

$$\phi(E_k(L)) \subseteq E_{k-1}(L_{\hat{\Omega}}) + I_{\mu-1}E_k(L_{\hat{\Omega}}). \quad //$$

Traldi gave a number of examples to show that this theorem is best possible, in the sense that either both inclusions could be strict, or one or both could be equalities.

Some Consequences of Torres' Conditions

The Torres conditions enable us to argue inductively about the first Alexander polynomial of a link from those of its sublinks. The following theorem is an example.

<u>Theorem 3</u> <u>If</u> L <u>is a</u> μ-<u>component link with</u> $\mu \geq 2$, <u>then</u>

$$\Delta_1(L) \text{ is in } I^{\mu-2} .$$

<u>Proof</u> It will suffice to show that all the partial derivatives of $\Delta_1(L)$ of order less than $\mu - 2$ vanish at $t_1 = \ldots = t_\mu = 1$. We shall induct on μ. There is nothing to prove in the 2-component case. Suppose the result holds for all links with fewer than μ components. No partial derivation $D_\alpha : f \longmapsto \partial^{|\alpha|}f/\partial t_1^{\alpha_1}\ldots\partial t_\mu^{\alpha_\mu}$ of order $|\alpha| = \Sigma_{i=1}^{\mu} \alpha_i$ less than μ can

involve differentiation with respect to all μ variables t_i. Suppose that D_α is such a partial derivation with $\alpha_\mu = 0$. The Torres conditions imply that there is some g in Λ_μ such that $\Delta_1(L) = (\Pi - 1)\Delta_1(L_{\hat{\mu}}) + (t_\mu - 1)g$ (where Π and $L_{\hat{\mu}}$ are as above).

Therefore

$D_\alpha\Delta_1(L) = \sum_{\beta\leq\alpha} (\alpha!/(\alpha-\beta)!\beta!)D_\beta(\Pi-1).D_{\alpha-\beta}\Delta_1(L_{\hat{\mu}}) + (t_\mu-1)D_\alpha g$ where the summation is taken over multi-indices $\beta = (\beta_1,\ldots,\beta_\mu)$ such that $\beta_i \leq \alpha_i$ for all i, and $\beta! = \prod_{1\leq i\leq\mu} \beta_i!$. Hence

$$\varepsilon(D_\alpha\Delta_1(L)) = \sum_{0<\beta\leq\alpha} (\alpha!/(\alpha-\beta)!\beta!)(\prod_{1\leq i<\mu} (\ell_{i\mu}!/(\ell_{i\mu}-\beta_i)!\beta_i!))\varepsilon(D_{\alpha-\beta}\Delta_1(L_{\hat{\mu}}))$$

If $|\alpha| < \mu - 2$ and $0 < \beta \leq \alpha$, then $|\alpha - \beta| < (\mu - 1) - 2$, so all the terms $\varepsilon(D_{\alpha-\beta}\Delta_1(L_{\hat{\mu}}))$ are 0, by the inductive assumption, and therefore $\varepsilon(D_\alpha(\Delta_1(L)) = 0$. Since a similar argument works for the other partial derivatives of order less than $\mu - 2$, the theorem is proven. //

If L is obtained from the 2-component link with abelian link group by replacing the second component by $\mu - 1$ parallel pairwise unlinked copies of itself, then $\Delta_1(L) = (t_1 - 1)^{\mu-2}$ and so this result is in general best possible. A similar argument shows that if for every $(\mu-1)$-component sublink $L_{\hat{i}}$ of L each partial derivative of $\Delta_1(L_{\hat{i}})$ of order less than $\mu - 2$ (respectively, $\mu - 1$) vanishes at $t_1 = \ldots = t_\mu = 1$, then all the partial derivatives of $\Delta_1(L)$ of order less than $\mu - 1$ (respectively, μ) vanish there. (Note that if $\mu \geq 2$, $\Delta_1(L)$ is in I^m if and only if $E_1(L) \subseteq I^{m+1}$). If all the linking numbers $\ell_{i\mu}$ are 0, the Torres conditions and the Remainder Theorem imply that $(t_\mu - 1)$ divides $\Delta_1(L)$. Therefore if $\ell_{ij} = 0$ for all i and j, the

Alexander polynomial is divisible by $\prod_{1\leq i\leq\mu}(t_i - 1)$ and so is in I^μ.

In his thesis Traldi used a similar argument by induction to show that if $k < \mu$ then $E_k(L) \subseteq I^{\mu-k} + (t_i - 1)$ for each $1 \leq i \leq \mu$, and hence that $E_k(L) \subseteq I^{\mu-k}$. He has since announced the following stronger result: "Let $M = [m_{ij}]$ be the $\mu \times \mu$ Λ_μ-matrix with entries $m_{ii} = 1 - \prod_{1\leq j\leq\mu} t_i^{\ell_{ij}}$ and, for $i \neq j$, $m_{ij} = t_i^{\ell_{ij}} - 1$. Then for $0 < k < \mu$

$$\sum_{0\leq i\leq\mu-k} E_{k+i}(L).I^{2i} + I^{2(\mu-k)} = \sum_{0\leq i\leq\mu-k} E_{k+i}(M).I^{2i} + I^{2(\mu-k)}.$$ In particular if $\ell_{ij} = 0$ for all i, j then $E_1(L) \subseteq I^{2(\mu-2)}$ and hence $\Delta_1(L)$ is in $I^{2\mu-3}$." (This is a paraphrase of part of his announcement of the paper [192]).

Theorem 3 says nothing of interest when $\mu = 2$. However the Torres conditions imply that $\varepsilon(\Delta_1(L)) = \pm \ell_{12}$ in this case. This is also a consequence of Milnor's theorem as the following proof of another result announced by Traldi shows.

<u>Theorem 4</u> (Traldi [191]) <u>Let</u> L <u>be a</u> μ-<u>component link, with</u> $\mu \geq 2$. <u>Let</u> d_i = h.c.f. $\{\ell_{ij} \mid 1 \leq j \leq \mu\}$ <u>for each</u> $1 \leq i \leq \mu$. <u>Then</u> $E_{\mu-1}(L) + I^2 = (d_1(t_1 - 1),\ldots d_\mu(t_\mu - 1)) + I^2$.

<u>Proof</u> Let $B = G'/G''$. Since $\mathbb{Z} \otimes A(L) = \mathbb{Z} \otimes I$, $B \subsetneq I.A(L)$ and so $IB \subsetneq I^2A(L)$. Therefore $A(L)/I^2A(L)$ depends only on G/G_3, since $B/IB \cong G'/G_3G'' = G_2/G_3$. By Milnor's Theorem G/G_3 has a presentation of the form

$$\{x_i \quad 1 \leq i \leq \mu \mid [x_i, \ell_i^{(3)}] = 1 \quad , [\ , [\ \ ,\ ]]\}$$

where $\ell_i^{(3)}$ represents an ith longitude modulo G_3. We may assume $\ell_i^{(3)} = x_1^{\ell_{i1}} \ldots x_\mu^{\ell_{i\mu}} w_i$ for some w_i representing an element of $G' = G_2$. Let $\partial_j : F(\mu) \to \Lambda$ be the Fox free derivative such that $\partial_j(x_j) = 1$ and $\partial_j(x_k) = 0$ if $j \neq k$. Then $E_{\mu-1}(G/G_3)$ is generated (modulo I^2) by $\{\partial_j[x_i, \ell_i^{(3)}] \mid 1 \leq i, j \leq \mu\}$. Since $\partial_i[x_i, \ell_i^{(3)}] = (t_i - 1)\ell_{ij} \cdot \prod_{1 \leq k \leq \mu} t_k^{\ell_{ik}}$ if $i \neq j$ and $\partial_j[x_j, \ell_j^{(3)}] = \left(\prod_{1 \leq k \leq \mu} t_k^{\ell_{jk}} - 1\right)$, which is congruent to $\sum_{1 \leq k \leq \mu} \ell_{jk}(t_k - 1)$ modulo I^2, and since $E_{\mu-1}(G/G_3) + I^2 = E_{\mu-1}(L) + I^2$, the theorem follows. //

The assertion of this theorem is equivalent to the case $k = \mu - 1$ in the above announcement of Traldi, who has advised us that he used the Milnor presentation of G/G_n also, to obtain his more general results. Traldi has also announced that $E_{\mu-1}(L) \subseteq J + D$, where $J = E_{\mu-1}(\bigoplus_{1 \leq i \leq \mu} \Lambda/(t_i - 1))$ is the ideal generated by the products $(t_i - 1)(t_j - 1)$ with $1 \leq i \neq j \leq \mu$, and where D is the ideal generated by the elements $t_k^{d_k} - 1$, for $1 \leq k \leq \mu$. (Notice that $D + I^2 = (d_1(t_1 - 1), \ldots, d_\mu(t_\mu - 1)) + I^2$). This assertion implies that of Theorem 4 also.

Insufficiency of the Torres Conditions

Seifert showed that an integral Laurent polynomial δ in Λ_1 which was symmetric of even degree ($\bar{\delta} = t^{2m}\delta$ for some m) and such that $\varepsilon(\delta) = \pm 1$ was the first Alexander polynomial of some knot [167]. His method was to embed a punctured surface in S^3 with prescribed self-linking characteristics. Using surgery on the knot complement, Levine showed that there was a knot for which $G'/G'' \approx \Lambda_1/(\delta)$, and by taking connected sums of knots with G'/G'' a cyclic Λ_1-module he obtained a similar characterization for the family of all Alexander polynomials of a knot [115].

The conditions $\bar{\delta} = t^{2m}\delta$ and $\varepsilon(\delta) = \pm 1$ are just the Torres conditions when $\mu = 1$. The obvious question is whether the Torres conditions suffice also when $\mu > 1$. (See also [50; Problem 2]). Levine has shown how to generate Alexander polynomials for links with prescribed linking matrix from a given one by surgery and hence has shown that if $\mu = 2$ and the linking number ℓ is ± 1 the Torres conditions suffice [117]. Bailey has characterized the Λ_1-module G'/G'' for a 2-component link in terms of the form of a presentation matrix, and hence has verified that they suffice also if $\ell = 0$ [7]. (We shall state Bailey's Theorem below). Kidwell has shown that for $\mu = 2$ and $\ell = 3$, and under restrictions on the "order" of the link, the Torres conditions are insufficient [99]. He has also shown that if $\mu \geq 3$ and all the linking numbers are 0, then the reduced Alexander polynomial $\Delta_1(L)(t, \ldots, t)$ must be highly divisible by $t - 1$ [100]. (See the next Chapter). However if $\mu > 2$, the Torres conditions do not always suffice to deduce the linking numbers, and so this result need not imply that they are insufficient.

In this section we shall assume that $\mu = 2$ and shall show that even then the Torres conditions are in general not sufficient. As they determine the linking number (up to sign) when $\mu = 2$, and as Bailey and Levine have settled the cases $\ell = 0$ or ± 1, we shall assume that $\ell > 1$. Our argument shall rely on the following theorem of Bailey [7; page 32. See also 144]:

"Theorem A Λ_2-module is a link module if it has a presentation matrix of the form

$$\begin{pmatrix} 1 - (xy)^{\ell}/1-xy & -(1-x)(1-y)(1 - (xy)^{\ell-1}/1 - xy)\beta(x,y) \\ \beta(x^{-1},y^{-1})^{tr} & \mathbb{A}(x,y) \end{pmatrix}$$

where $\mathbb{A}(x,y)$ is a square matrix, $\beta(x,y)$ is a row matrix, both with entries in Λ_2, satisfying $\mathbb{A}(x,y) = \mathbb{A}(x^{-1}, y^{-1})^{tr}$ and $\mathbb{A}(1,1) = \mathrm{diag}(\pm 1, \ldots, \pm 1)$. Furthermore $\mathbb{A}(x,1)$ (respectively, $\mathbb{A}(1,y)$) is a presentation matrix for the first (respectively, second) component of the link and ℓ is the linking number of the two components."

The entry in the (1,1)-position of the above matrix is the first Alexander polynomial of the link which bounds an annulus embedded in S^3 with unknotted core and ℓ full twists. Bailey proved his theorem by observing that any 2-component link with linking number ℓ could be obtained by surgery on this link, and by using Alexander duality to compute a presentation matrix for the link module. (Here the <u>link module</u> is the Λ_2-module G'/G''. We shall follow Bailey in using x and y instead of t_1 and t_2 for our Laurent polynomial variables).

One corollary of Bailey's theorem is that the Alexander polynomial of a 2-component link has the form

$$\Delta(x,y) = (1 - (xy)^{\ell}/1 - xy).A(x,y) - (1 - x)(1 - y)(1 - (xy)^{\ell-1}/1 - xy).B(x,y)$$

with $A(x,y) = A(x^{-1},y^{-1})$, $B(x,y) = B(x^{-1},y^{-1})$ and $A(x,1)$ and $A(1,y)$ knot

polynomials. (For instance we may take $A(x,y) = \det A(x,y)$,
$B(x,y) = \det \begin{pmatrix} 0 & \beta \\ \bar{\beta}^{tr} & A \end{pmatrix}$.) He showed moreover that a polynomial in Λ_2 has this form if and only if it satisfies both Torres conditions. Given a polynomial satisfying the Torres conditions, its expression in the above form is not unique. However if A, B and A', B' both give rise to Δ, then there is some $C \in \Lambda_2$ such that $A - A' = (1 - x)(1 - y)(1 - (xy)^{\ell-1}/1 - xy).C$ and $B - B' = (1 - (xy)^{\ell}/1 - xy).C$. We may therefore remove the ambiguity by passing to a quotient ring in which each of $(1 - x)(1 - y)(1 - xy)^{\ell-1}/1 - xy)$ and $(1 - (xy)^{\ell}/1 - xy)$ are mapped to 0. If $\wp$ is a proper prime ideal of Λ_2 containing these two elements, then it must contain either $1 - x$ or $1 - y$, since it is prime and since $(1 - (xy)^{\ell-1}/1 - xy)$ and $(1 - (xy)^{\ell}/1 - xy)$ together generate the unit ideal. If $1 - y \in \wp$ then $(1 - x^{\ell}/1 - x) \in \wp$ so $\wp$ must contain the d-cyclotomic polynomial $\Phi_d(x)$ for some $d > 1$ dividing ℓ. (Similarly if $1 - x \in \wp$, then $\wp$ must contain $\Phi_e(y)$ for some $e > 1$ dividing ℓ.) Thus if we wish our quotient ring to be an integral domain, it is no loss of generality to fix a primitive d^{th} root of unity ζ and to consider the homomorphism $f : \Lambda_2 \longrightarrow R = \mathbb{Z}[\zeta]$ mapping x to ζ and y to 1. Notice that the involution $x \to x^{-1}$, $y \to y^{-1}$ of Λ_2 gives rise to complex conjugation $\zeta \to \bar{\zeta} = \zeta^{-1}$ on the quotient ring $R = \Lambda_2/\wp$.

The images f(A) and f(B) of A and B in R are well defined and so we may ask how they may be determined from Δ. The map f factors through the projection of Λ_2 onto $\Lambda_1 = \Lambda_2/(1 - y)$, and we see that f(A) is the value of the knot polynomial $(x - 1)(x^{\ell} - 1)^{-1}\Delta(x,1)$ at $x = \zeta$. On considering the Taylor expansion of Δ, A, B etc. about $(x,y) = (1,\zeta)$ we see that f(A) is also equal to $\ell^{-1}.\zeta(\zeta - 1).\partial\Delta/\partial x(\zeta,1)$, and that $f(B) = B(\zeta,1) = \zeta.(\zeta - 1)^{-1}.(\partial\Delta/\partial y(\zeta,1) - \zeta.\partial\Delta/\partial x(\zeta,1))$. It is easily checked that f(A) and f(B) are real, using the Torres conditions and the fact that complex conjugation is induced by the involution of Λ_2. Since the coefficients of Δ are rational, whether $f(A) = 0$ does not depend on the choice of the primitive d^{th} root ζ.

We shall now state and prove our main result

Theorem 5 Let L be a 2-component link with linking number $\ell > 1$, and with first Alexander polynomial $\Delta(x,y)$. If the knot polynomial $(x-1/x^{\ell} -1)\Delta(x,1)$ is (up to a unit) the d-cyclotomic polynomial $\Phi_d(x)$ for some $d > 1$ dividing ℓ, and if ζ is a primitive d^{th} root of unity, then the $\mathbb{Z}[\zeta]$-ideal generated by $\zeta(\zeta - 1)^{-1}.\partial\Delta/\partial y(\zeta,1)$ is of the form $J\bar{J}$ for some ideal J.

Proof By Bailey's theorem there are square matrices $\mathbb{A}$ and $\mathbb{B} = \begin{pmatrix} 0 & \beta \\ \bar{\beta}^{tr} & \mathbb{A} \end{pmatrix}$ which are Hermitean with respect to the involution of Λ_2 and such that $\Delta = (1 - (xy)^{\ell}/1 - xy).\det \mathbb{A} - (1 - x)(1 - y)(1 - (xy)^{\ell-1}/1 - xy).\det \mathbb{B}$. If $\Phi_d(x)$ divides $(x - 1)(x^{\ell} -1)^{-1}\Delta(x,1) = \det \mathbb{A}(x,1)$, then $\det f(\mathbb{A}) = f(\det \mathbb{A}) = 0$.

Suppose first that $R = \mathbb{Z}[\zeta]$ is a Euclidean domain. Then we may reduce one of the rows of $f(\mathbb{A})$ to 0 by elementary row operations. Since any elementary R-matrix may be lifted to an elementary Λ_2-matrix, we may thus find a Λ_2-matrix $\mathbb{P}$ with determinant 1 such that $f(\mathbb{P}\,\mathbb{A}\,\bar{\mathbb{P}}^{tr})$ has first row and column 0. (We perform the conjugate column operations also so as to preserve the Hermitean character of the matrices.) Therefore if $\mathbb{Q} = 1 \oplus \mathbb{P}$ then $\det \mathbb{B} = \det(\mathbb{Q}\,\mathbb{B}\,\bar{\mathbb{Q}}^{tr})$ and $\mathbb{Q}\,\mathbb{B}\,\bar{\mathbb{Q}}^{tr}$ has the form

$$\begin{pmatrix} 0 & \beta_1 & \gamma \\ \bar{\beta}_1 & a\Phi_d(x) + b(y-1) & \Phi_d(x)\mu + (y-1)\nu \\ \bar{\gamma}^{tr} & \Phi_d(x^{-1})\bar{\mu}^{tr} + (y^{-1}-1)\nu & \mathbb{C} \end{pmatrix}$$

for some square matrix $\mathbb{C} = \bar{\mathbb{C}}^{tr}$, row matrices γ, μ and ν, and elements $a = \bar{a}$, $b = \bar{b}$, β_1 of Λ_2. Then $f(\det \mathbb{B}) = f(\det(\mathbb{Q}\,\mathbb{B}\,\bar{\mathbb{Q}}^{tr})) = -\,f(\beta_1)\overline{f(\beta_1)}f(\det \mathbb{C})$, and $f(\Phi_d(x)^{-1}\det \mathbb{A}(x,1)) = f(\Phi_d(x)^{-1}\det(\mathbb{P}\mathbb{A}(x,1)\bar{\mathbb{P}}^{tr})) = f(a).f(\det \mathbb{C})$. Therefore if $\det \mathbb{A}(x,1) = \Phi_d(x)$, $f(\det \mathbb{C})$ is a (real) unit and so the ideal generated by $\zeta.(\zeta - 1)^{-1}.\partial\Delta/\partial y(\zeta,1) = f(\det \mathbb{B})$ equals $J\bar{J}$, where J is the ideal generated by $f(\beta_1)$.

The above argument is directly applicable for only finitely many d [124], but the general case may be recovered by localization, since R is a Dedekind domain. Let $\mathfrak{q}$ be a prime ideal of R, and for each ideal I of R or $R_{\mathfrak{q}}$ define $V_{\mathfrak{q}}(I)$ by the equation $I_{\mathfrak{q}} = IR_{\mathfrak{q}} = \mathfrak{q}^{V_{\mathfrak{q}}(I)}R_{\mathfrak{q}}$. The argument of the previous paragraph (with $\Lambda_{2,f^{-1}(\mathfrak{q})}$ and $R_{\mathfrak{q}}$ in place of Λ_2 and R) shows that $(f(\det \mathbb{B})_{\mathfrak{q}} = (f(b_{\mathfrak{q}})\overline{f(b_{\mathfrak{q}})})$ for some $b_{\mathfrak{q}}$, and hence that if $\mathfrak{q} = \bar{\mathfrak{q}}$, $V_{\mathfrak{q}}((f(\det \mathbb{B}))) = 2V_{\mathfrak{q}}((f(b_{\mathfrak{q}}))) = 2W_{\mathfrak{q}}$ say. If $\mathfrak{q} \neq \bar{\mathfrak{q}}$, then $V_{\mathfrak{q}}((f(\det \mathbb{B}))) = V_{\bar{\mathfrak{q}}}((f(\det \mathbb{B}) = Z_{\mathfrak{q}}$ say, since $f(\det \mathbb{B})$ is real. Let $S = \{\mathfrak{q} \neq \bar{\mathfrak{q}} \mid Z_{\mathfrak{q}} > 0\}$, and let $T \subset S$ contain exactly one representative of each complex conjugate pair. Let $J = (\prod_{\mathfrak{q}=\bar{\mathfrak{q}}} \mathfrak{q}^{W_{\mathfrak{q}}}).(\prod_{\mathfrak{r}\in T} \mathfrak{r}^{Z_{\mathfrak{r}}})$. Then $V_{\mathfrak{q}}((f(\det \mathbb{B}))) = V_{\mathfrak{q}}(J\bar{J})$ for all primes $\mathfrak{q}$ of R, so we may conclude that $(f(\det \mathbb{B})) = J\bar{J}$ [168; page 23]. This proves the theorem. //

The hypothesis of the theorem is vacuous unless d is divisible by at least 2 primes, for $\Phi_d(1)$ must divide a knot polynomial, and so must be ± 1. Therefore the first case to look at corresponds to $\ell = 6$, and then $\Phi_6(x) = x^2 - x + 1$. Consider the polynomial

$$D(x,y) = (1-(xy)^6/1-xy).(x-1+x^{-1}) - (1-x)(1-y)(1-(xy)^5/1-xy).2.$$

Then it is easily verified that D satisfies the conditions of Torres, but that if ω is a primitive 6th root of unity, then $\omega.(\omega-1)^{-1}\,\partial D/\partial y(\omega,1) = 2$, and the $\mathbb{Z}[\omega]$-ideal generated by 2 is clearly not of the form $J\bar{J}$. Thus D cannot be the first Alexander polynomial of any 2-component link. Notice that in this case the ring $\mathbb{Z}[\omega]$ is actually an Euclidean domain [124] (This example, and subsequently the above theorem, was suggested by the question in Bailey's thesis [7; page 69] on whether there were any matrices $\mathbb{A}$, $\mathbb{B}$ as above such that $(\det \mathbb{A}, \det \mathbb{B}) = (x-1+x^{-1}, 2)$. The argument of the theorem extends readily to give a necessary condition for a pair of elements $(a,b) \in \Lambda_2^2$ to be of the form $(\det\mathbb{A}, \det \mathbb{B})$. Suppose $\alpha \in \Lambda_2$ is a simple prime factor of a such that $(\bar{\alpha}) = (\alpha)$, and that $\wp$ is a prime ideal containing α such that $\wp = \bar{\wp}$ and $R = \Lambda_2/\wp$ is Dedekind. Then the ideal generated by the image of b in R must be of the form $cJ\bar{J}$ for some ideal J and some element c which divides the image of $\alpha^{-1}a$).

Murasugi's Theorem

Although the above theorem follows almost inevitably from Bailey's Theorem, its meaning is still rather obscure. The cyclotomic polynomials surely suggest that the homology of a d-fold cyclic cover of the link exterior is involved. The role of the partial derivative $\partial\Delta/\partial y(\zeta,1)$ is not at all clear. In this section we shall sketch a proof of a theorem of Murasugi which shows that the ideal generated by this derivative in $\mathbb{Z}[\zeta]$ is invariant under homotopy of the second component of the link. We are grateful to Murasugi for sending us an outline in English of his theorem, which has only been published in Japanese [142a,b].

Theorem (Murasugi) Let L^+ and L^- be μ-component links which share the same $(\mu-1)$-component sublink K obtained by deleting the μth components, and such that L^+_μ is homotopic to L^-_μ in $S^3 - K$. Let Δ^+ and Δ^- be the first Alexander polynomials of L^+ and L^- respectively. Then $\frac{\partial}{\partial t_\mu}\Big|_{t_\mu=1}\Delta^+$ and $\frac{\partial}{\partial t_\mu}\Big|_{t_\mu=1}\Delta^-$ generate the same ideal in $\Lambda_{\mu-1}/(\Delta_1(K))$.

Proof As the argument is no different in the general case, we shall assume that $\mu = 2$ and write x,y for t_1,t_2 respectively. It will suffice to assume also that L^- is obtained from L^+ by changing one overcrossing of L^+ to an undercrossing. Thus we may depict the two links as in Figure 1. Murasugi's idea is to compare Δ^+ and Δ^- with the first Alexander polynomial Δ^0 of the $(\mu+1)$-component link L^0 depicted in the figure. He deduces the theorem on applying the second Torres condition and the following lemma.

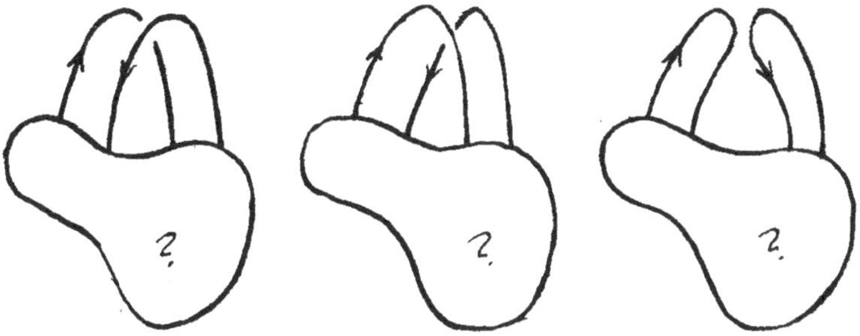

L^+ L^- L^o

Figure 1

<u>Lemma</u> <u>For suitable choices of the Alexander polynomials,</u>

$$\partial\Delta^+/\partial y(x,1) + \partial\Delta^-/\partial y(x,1) = \Delta^o(x,1,1).$$

This lemma is a straightforward computation, based on the fact proven in [40, 139, 189] that the determinant obtained from the Jacobian matrix of a Wirtinger presentation of a link group by deleting a column corresponding to an ith meridian and any one row is $(t_i - 1)$ times the first Alexander polynomial of the link. The lemma may also be proven by differentiating the first Conway identity [32, 86]:

$$\nabla_+ - \nabla_- = (y^{\frac{1}{2}} - y^{-\frac{1}{2}})\nabla_o.$$

Neither approach is in the spirit of the rest of these notes. As our Theorem 5 depends on Bailey's use of surgery to change the crossings of components of a link with themselves, in other words to carry out a link homotopy, we might expect a deeper connection between it and Murasugi's Theorem, and in return we should seek a proof of the latter theorem via surgery.

CHAPTER VIII REDUCED ALEXANDER IDEALS

The methods of the above chapters may also be applied to the homology of other covering spaces of link complements, and in particular to the infinite cyclic cover determined by the total linking number homomorphism. For knots this is the maximal abelian cover, and may be constructed by splitting along a Seifert surface. As this technique works for any link, the total linking number cover has been studied extensively. (See for instance [84, 100, 138, 172, 182]). It is of particular interest when the link complement fibres over the circle; such is the case for the links associated with algebraic singularities [133], and here the reduced Alexander polynomial is the characteristic polynomial of the monodromy.

In this chapter weshall give "coordinate free" proofs of results of Hosokawa and Kidwell on the divisibility of the reduced Alexander polynomial of a link. The conditions of Torres and Traldi are used to show that the Hosokawa polynomial $\nabla(L)$ is symmetric of even degree, and to evaluate the integer $|\nabla(L)(1)|$. We consider links for which the rank of the reduced Alexander module is maximal, and prove once again that the link of Figure V.1 is not an homology boundary link. Next we define fibred links, and show that the reduced Alexander polynomial of a fibred 1-link is nonzero, while the only fibred links in higher dimensions are fibred knots. In this section we list without proof some of the properties of the monodromy of algebraic links. We conclude the chapter with a brief summary of some results on the branched coverings of a link.

The Total Linking Number Cover

Let L be a μ-component n-link with exterior X, pointed by *, and group G, and let $\tau : G \to \mathbb{Z}$ be the unique homomorphism sending each meridian to 1. This determines a homomorphism from $\mathbb{Z}[G/G']$ to $\mathbb{Z}[\mathbb{Z}]$ which corresponds to the projection $\pi : \Lambda_\mu \to \Lambda = \mathbb{Z}[t,t^{-1}]$ sending each variable t_i to t. (Throughout this chapter Λ shall denote $\mathbb{Z}[t,t^{-1}]$ only.)

Definition The total linking number cover of X is the cover $q : X^\tau \to X$ determined by $\ker \tau$.

It is readily verified that a loop in X lifts to a loop in X^τ if and only if the sum of its linking numbers with various (oriented) components of L in S^{n+2} is 0; whence the name "total linking number cover". The Cartan-Leray spectral sequence for the cover q reduces to the Wang sequence of the fibration $X^\tau \to X \xrightarrow{\tilde{\tau}} S^1$ (where $\tilde{\tau}$ represents the unique homotopy class of maps inducing τ), and we shall invoke it as the "Wang sequence for q".

The total linking number cover of an n-link is considered by Sumners in [182], where he relates the $\mathbb{Z}$- and Λ-module structures on the homology of X^τ, extending results of Crowell for knots [38], and gives necessary and sufficient conditions for an n-knot with group $\mathbb{Z}$ to fibre over the circle. Shinohara and Sumners [172] study the rational homology of X as a $Q\Lambda$-module, extending results of Levine for knots [116], and deduce criteria for the link to be splittable. We shall however not consider the high dimensional case in detail, and in the next three sections shall assume that n = 1.

The Hosokawa Polynomial

If L is a 1-link the equivariant chain complex of $(X^{\tau}, q^{-1}(*))$ is chain homotopy equivalent to one of the form

$$0 \longrightarrow \Lambda^a \xrightarrow{\pi(d)} \Lambda^{a+1} \longrightarrow 0 \longrightarrow 0$$

where $d : \Lambda_\mu^a \rightarrow \Lambda_\mu^{a+1}$ is the boundary map for the corresponding complex for $(X', p^{-1}(*))$. Thus $H_1(X,*;\Lambda) = H_1(X^{\tau}, q^{-1}(*);\mathbb{Z}) = \Lambda \otimes_{\Lambda_\mu} A(L)$, which we shall abbreviate as $\pi A(L)$. (This may be called the reduced Alexander module of L). Since $\tilde{H}_0(*;\Lambda) \approx I_1 \approx \Lambda$, and $H_1(*,\Lambda) = 0$, there is an exact sequence

$$0 \longrightarrow H_1(X;\Lambda) \longrightarrow \pi A(L) \longrightarrow \Lambda \longrightarrow 0$$

so $\pi A(L) \approx \Lambda \oplus H_1(X;\Lambda)$. Hence $E_i(H_1(X;\Lambda)) = E_{i+1}(\pi A(L)) = \pi(E_{i+1}(L))$ for all i. In particular $E_0(H_1(X;\Lambda)) = \pi(E_1(L))$ is principal, by Theorem IV.3.

<u>Definition</u> The <u>reduced Alexander polynomial</u> of L is

$$\Delta_{red}(L)(t) = \pi(\Delta_1(L)) = \Delta_1(L)(t,\ldots,t).$$

By Theorem VII.3 $\Delta_1(L)$ is in $I_\mu^{\mu-2}$, so $\pi(\Delta_1(L))$ is in $(\pi(I_\mu))^{\mu-2} = (t-1)^{\mu-2}$ and so $\pi(E_1(L)) \subseteq (t-1)^{\mu-1}$ (using Theorem IV.3 again, if $\mu > 1$). Therefore the following definition is possible.

<u>Definition</u> The <u>Hosokawa polynomial</u> of L is the generator $\nabla(L)(t)$ of the principal ideal $(t-1)^{1-\mu}E_1(\pi A(L))$ which satisfies

$\nabla(L)(t) = \Delta_1(L)(t)$ if $\mu = 1$ and $\nabla(L)(t) = \Delta_{red}(L)(t)/(t-1)^{\mu-2}$ if $\mu > 1$.

Of course this definition depends on a choice of representatives for the first Alexander polynomial, but the ambiguity shall be quite harmless here. We shall usually abbreviate $\Delta_{red}(L)(t)$ and $\nabla(L)(t)$ by Δ_{red} and ∇. Note that $\Delta_{red} \neq 0$ if and only if $H_2(X;\Lambda) = H_2(X^{\tau};\mathbb{Z})$ is 0.

Hosokawa proved that Δ_{red} was divisible by $(t - 1)^{\mu-2}$ by computing linking numbers of cycles on a Seifert surface [84]. (We shall give yet another proof shortly by means of the Wang sequence for q). The example following Theorem VII.3 shows that in general Δ_{red} need not be contained in $(t - 1)^{\mu-1}$, in other words that $\varepsilon(\nabla)$ need not be 0.

Hosokawa showed that ∇ is symmetric of even degree ($\bar{\nabla} = t^{2m}\nabla$ for some m), and that any such symmetric polynomial was the Hosokawa polynomial of a μ-component link, for each $\mu > 1$. (Thus the ambiguity in the definition of ∇ may be reduced to one of sign, by replacing ∇ by $t^m\nabla$, so that $\overline{t^m\nabla} = t^m\nabla$). Furthermore he computed the absolute value of the integer $\varepsilon(\nabla)$ as a determinant in the linking numbers of L.

If $\mu = 1$, the Hosokawa polynomial is the first Alexander polynomial of the knot, and it was shown in Chapter VII to be symmetric of even degree and to augment to ± 1. Seifert showed that any such polynomial is the Alexander polynomial of a knot [167].

If $\mu > 1$, we may derive the symmetry conditions from the first Torres conditions, that $\overline{\Delta_1(L)} = (-1)^{\mu}\left(\prod_{1 \leqslant i \leqslant \mu} t_i^{b_i}\right)\Delta_1(L)$ where

$b_i \equiv 1 - \sum_{1 \leqslant j \leqslant \mu} \ell_{ij}$ modulo (2). For then

$(\bar{t}-1)^{\mu-2}\bar{\nabla} = (-1)^{\mu} t^{\Sigma b_i}(t-1)^{\mu-2}\nabla$, so $\bar{\nabla} = t^{b}\nabla$ where

$b = 2-\mu+\sum_{1 \leqslant i \leqslant \mu} b_i \equiv \sum_{1 \leqslant i,j \leqslant \mu} \ell_{ij}$ modulo (2) is even.

Following Traldi, we may deduce the value of $|\varepsilon(\nabla)|$ from the equality of his announcement [192]. We shall first prove a lemma needed below, which shall also provide another proof that $(t-1)^{\mu-2}$ divides $\Delta_1(L)$.

Lemma 1 Let W be a finite cell complex with $H_1(W;\mathbb{Z}) \approx \mathbb{Z}^{\mu}$, and let $p : W^{\infty} \to W$ be a connected infinite cyclic cover. Let $H = H_1(W^{\infty};\mathbb{Z})$ considered as a Λ-module via the action of a fixed generator for the covering group. Then

$$E_i(H) \subseteq (t-1)^{\mu-1-i} \quad \text{for} \quad i < \mu .$$

Proof It will suffice to show for each $i < \mu$ that the ideal generated by $(t-1)^{\mu-1-i}$ in $Q\Lambda = Q[t,t^{-1}]$ contains that generated by the image of $E_i(H)$. The latter ideal is just $E_i(H \otimes Q)$, and $H \otimes Q$ is the first homology of W^{∞} with rational coefficients, considered as a $Q\Lambda$-module. Since $Q\Lambda$ is a principal ideal domain, we may write $H \otimes Q = \bigoplus_{i \leqslant s \leqslant r} (Q\Lambda/(\theta_s))$

where θ_{s+1} divides θ_s in $Q\Lambda$ for $1 \leqslant s < r$. The Wang sequence (with rational coefficients) for the cover p

$$\ldots \longrightarrow H(W^{\infty};Q) \xrightarrow{t-1} H_1(W^{\infty};Q) \longrightarrow H_1(W;Q) \longrightarrow Q \xrightarrow{0} Q \longrightarrow Q$$

then shows that $H \otimes Q/(t-1)H \otimes Q = \bigoplus_{1 \leqslant s \leqslant r} (Q/\theta_s(1).Q)$ is isomorphic to $Q^{\mu-1}$. Hence the first $\mu - 1$ of the numbers $\theta_s(1)$ must be 0 and so $E_i(H \otimes Q) = \left(\prod_{i+1 \leqslant s \leqslant r} \theta_s\right)$ is contained in $(t-1)^{\mu-1-i}$. //

If L is a μ-component link we may take W = X in the above lemma and conclude that $\pi(E_j(L)) = E_{j-1}(H_1(X;\Lambda))$ is contained in $(t-1)^{\mu-j}$ for $1 \leqslant j \leqslant \mu$. (As remarked in Chapter VII it is in fact true that $E_j(L) \subseteq I_\mu^{\mu-j}$ for $j \leqslant \mu$.) On applying this lemma to Traldi's equality in the case $k = 1$, we see that $\pi(E_1(L)) \equiv \pi(E_1(M))$ modulo $(t-1)^\mu$. (In fact $E_1(L) \equiv E_1(M)$ modulo I_μ^μ). Now

$\pi(E_1(L)) = (t-1)^{\mu-1}(\nabla) \equiv (t-1)^{\mu-1}(\varepsilon(\nabla))$ modulo $(t-1)^\mu$, while

$\pi(m_{ii}) \equiv \left(-\sum_{1 \leqslant j \leqslant \mu} \ell_{ij}\right)\left(t-1\right)$ modulo $(t-1)^2$ and $\pi(m_{ij}) \equiv \ell_{ij}(t-1)$ modulo $(t-1)^2$ if $i \neq j$, so $\pi(E_1(M)) \equiv E_1(N).(t-1)^{\mu-1}$ modulo $(t-1)^\mu$ where $N = [n_{ij}]$ is the $\mu \times \mu$ $\mathbb{Z}$-matrix with $n_{ii} = -\sum_{1 \leqslant j \leqslant \mu} \ell_{ij}$ and $n_{ij} = \ell_{ij}$ if $i \neq j$. Hence $\varepsilon(\nabla)$ and $E_1(N)$ generate the same $\mathbb{Z}$-ideals, in other words $|\varepsilon(\nabla)| = |E_1(N)|$. (Note that since the sum of all the rows and the sum of all the columns of N are each 0, all the $(\mu-1) \times (\mu-1)$ minors of N have the same absolute value.)

If all the linking numbers are nonzero this last result may also be established via the Torres conditions, starting from

$$\varepsilon(\nabla) = \lim_{t \to 1} \Delta_{red}/(t-1)^{\mu-2} = \frac{1}{(\mu-2)!}\varepsilon(d^{\mu-2}\Delta_{red}/dt^{\mu-2})$$

$$= \sum_{|\alpha| = \mu-2} \delta_\alpha(\Delta_1(L))$$

where $\delta_\alpha f = \frac{1}{\alpha!}\,\varepsilon(\partial^{|\alpha|} f/\partial t_1^{\alpha_1} \ldots \partial t_\mu^{\alpha_\mu})$ for any f in Λ and any multi-index $\alpha = (\alpha_1,\ldots,\alpha_\mu)$. The terms $\delta_\alpha(\Delta_1(L))$ with $|\alpha| = \mu - 2$ may be computed inductively using the Torres conditions, and compared with the terms in the expansion of a principal $(\mu-1)\times(\mu-1)$ minor of N. The assumption on the linking number is needed to keep track of the ambiguity in sign of the Alexander polynomials of the sublinks.

In [89], Kawauchi has defined Hosokawa polynomials for more general infinite cyclic coverings, and has used an extension of Milnor duality to prove that these polynomials are of even degree and are symmetric up to sign.

Kidwell's Theorem

In this section we shall show that the divisibility of Δ_{red} by powers of $t-1$ can be more than doubled if all the linking numbers are 0. The following theorem was proven by Kidwell by a refinement of Hosokawa's approach, using a matrix derived from a connected spanning surface for the link. We shall assume that $\Delta_{red} \neq 0$.

Theorem 2 (Kidwell [100]) Let L be a μ-component link such that any two components of L have mutual linking number 0. Then $\Delta_{red} = (t-1)^{2\mu-3} f(t)$ for some f(t) in Λ such that $f(1) = 0$, if μ is even.

Proof Let $W = X \cup \mu(D^2 \times S^1)$ be the closed 3-manifold obtained by 0-framed surgery on the link L. Since all the linking numbers of L are 0, the inclusion map $H_1(X;\mathbb{Z}) \to H_1(W;\mathbb{Z})$ is an isomorphism, and so the total linking number cover extends to an infinite cyclic cover $r : W^\tau \to W$. By Lemma 1, $E_0(H_1(W^\tau;Q))$ is contained in $(t-1)^{\mu-1}$. The long exact sequence of the pair (W^τ, X^τ) gives an exact sequence

$$0 \to H_2(W^\tau;Q) \to H_2(W^\tau,W^\tau : Q) \to H_1(X^\tau : Q) \to H_1(W^\tau : Q) \to 0$$

(in which the first map is injective since $\Delta_{red} \neq 0$ implies that $H_2(X^\tau,Q) = H_2(X^\tau;\mathbb{Z}) \otimes Q = 0$). Therefore $H_1(W^\tau;Q)$ is a $Q\Lambda$-torsion module, so $H_2(W^\tau;Q) = H_2(W;Q\Lambda) \approx \overline{H^1(W;Q\Lambda)} \approx Ext_{Q\Lambda}(H_0(W;Q\Lambda),Q\Lambda) \approx Q = Q\Lambda/(t-1)$ by duality and Universal Coefficients. By excision, the second term is isomorphic to $H_2(r^{-1}(D^2 \times S^1), r^{-1}(S^1 \times S^1);Q) \approx (Q\Lambda/(t-1)^\mu)$. Therefore $E_0(H_1(W^\tau;Q)).(t-1)^\mu = E_0(H_1(X^\tau;Q)).(t-1) = (\Delta_{red}(t-1)^2)$, so $(t-1)^{2\mu-3} = (t-1)^{\mu-1+\mu-2}$ divides Δ_{red} in $Q\Lambda$, and therefore also in Λ. (Notice also that we have shown $E_0(H_1(W^\tau;Q)) = (\nabla)$.)

Let $f(t) = \Delta_{red}/(t-1)^{2\mu-3} = \nabla/(t-1)^{\mu-1}$. Since $\bar{\nabla} = t^{2m}\nabla$ for some m, $\overline{f(t)} = (-1)^{\mu-1}\, t^{2m+\mu-1} f(t)$ and so $f(1) = (-1)^{\mu-1} f(1)$. Therefore if μ is even, $f(1) = 0$. //

Kidwell proved that the integer $|f(1)|$ is the determinant of a $(\mu-1) \times (\mu-1)$ skew symmetric matrix derived from Hosokawa's matrix, and so is 0 if μ is even and is a perfect square if μ is odd.

Which elements of Λ may occur as $\nabla(L)/(t-1)^{\mu-1}$ for a μ-component link L all of whose linking numbers are 0?

If all the linking numbers of a μ-component link are 0, and if $\mu > 3$, then Kidwell's theorem implies that $\Delta_1(L) \Big/ \prod_{i-1}^{\mu} (t_i - 1)$ is in I, and so $\Delta_1(L)$ is in $I^{\mu+1}$. Can this exponent be improved upon? Traldi has announced that $\Delta_1(L)$ is in $I^{2\mu-3}$. ([192]. See Chapter VII). We have not been able to find a 4-component link, all of whose linking numbers are 0, for which the first Alexander polynomial is not in the 6th power of the augmentation ideal.

Null Reduced Ideals

Definition The reduced nullity of L is

$$\kappa(L) = \min\{k \mid \pi(E_k(L)) \neq 0\}.$$

It is obvious that $1 \leq \alpha(L) \leq \kappa(L) \leq \mu$ and it is easily checked that $\kappa(L) = 1 + \mathrm{rank}_\Lambda H_1(X^\tau; \mathbb{Z})$.

Theorem 3 The following are equivalent:

(i) $\kappa(L) = \mu$;

(ii) the longitudes of L are in $(\ker\tau)'$, the commutator subgroup of $\ker\tau$;

(iii) $H_1(X;\mathbb{Z}) \approx H_1(W;\mathbb{Z})$ and $H_1(X^\tau;\mathbb{Z}) \approx H_1(W^\tau;\mathbb{Z})$, where W is as in Theorem 2 above.

Proof If $\tilde{L}$ is a ν-component sublink of L, then $\kappa(L) - \kappa(\tilde{L}) \leq \mu - \nu$; hence $\kappa(\tilde{L}) = \nu$ if $\kappa(L) = \mu$. Since the Torres conditions imply immediately that a 2-component link with reduced nullity 2 has linking number 0, so all the longitudes of L are in $G' \subset \ker\tau$. Let ℓ_i be the image of the ith longitude in $H_1(X^\tau;\mathbb{Z}) \approx \ker\tau/(\ker\tau)'$. Then $(t-1).\ell_i = 0$ since each longitude commutes with a meridian. By an easy argument as in Theorem VI.1 there is a $\delta \in \pi(E_\mu(L))$ such that $\varepsilon(\delta) = 1$ and which annihilates the Λ-torsion submodule of $H_1(X^\tau;\mathbb{Z})$; hence $\mathrm{Ann}(\ell_i) \supseteq (t-1,\delta) = \Lambda$. Therefore all the longitudes are in $(\ker\tau)'$.

Conversely, all the longitudes are in G' if and only if $H_1(W;\mathbb{Z}) = \mathbb{Z}^\mu$, in which case there is an exact sequence

$$0 \to H_2(X^\tau;\mathbb{Z}) \to H_2(W^\tau;\mathbb{Z}) \to H_2(W^\tau,X^\tau;\mathbb{Z}) \xrightarrow{\partial} H_1(X^\tau;\mathbb{Z}) \to H_1(W^\tau;\mathbb{Z}) \to 0.$$

By excision, the middle Λ-module is isomorphic to $\mathbb{Z}^\mu$, and the boundary map ∂ is trivial if and only if either (ii) or (iii) is true. Thus (ii) and

(iii) are equivalent. The remaining implication now follows on taking rational coefficients Q and appealing to duality, which implies that $H_2(W^\tau;Q) \approx Q \oplus (Q\Lambda)^r$ where r is the rank over $Q\Lambda$ of $H_1(W^\tau;Q) \approx H_1(X^\tau;Q)$ and so equal to $\kappa(L) - 1$. //

By an argument using the Universal Coefficient spectral sequence and Poincaré duality, again as in Theorem VI.1, it may be shown that (if $\kappa(L) = \mu$) the ideal $\pi(E_\mu(L))$ is principal if and only if the Λ-module $H_1(X^\tau;\mathbb{Z}) \approx H_1(W^\tau;\mathbb{Z})$ has a free summand of rank $\mu-1$. This is always the case for an homology boundary link. More precisely, if $A(L)$ maps onto Λ^μ, then $\pi(E_\mu(L)) = (\pi\Delta_\mu(L))$. For $E_\mu(L) = (\Delta_\mu(L))E_0(B)$ by Theorem VI.1, and $\pi(E_\mu(L))$ is a principal ideal (by the above remark) which is not contained in $(t-1)$. Therefore the same is true of $\pi(E_0(B))$. But $\prod_{1 \leq i \leq \mu} (t_i - 1)$ annihilates B, so $\pi(E_0(B))$ contains $(t-1)^N$ for N large. Therefore $\pi(E_0(B)) = (1)$. However for the 2-component link $L = L_1 \cup L_2$ of Figure V.1, $\pi(E_2(L)) = (3, 1+t)^2$, and so this link is not an homology boundary link.

Fibred Links

Definition A μ-component n-link L is fibred if there is a fibre bundle projection $\phi : X \to S^1$ such that the induced map of fundamental groups is the total linking number homomorphism.

If $\mu = 1$, the final clause is redundant, and each fibre of ϕ is a Seifert surface for the knot. If $n = 1$, the example of the link  which has exterior homeomorphic to $S^1 \times S^1 \times [0,1]$ shows that the exterior may fibre over the circle in many ways, although there is essentially only one fibration satisfying the condition on meridians. We shall show below that when n is greater than 1 there are no fibred n-links with more than one component.

Let $e : \mathbb{R} \to S^1$ be the exponential map, sending r in $\mathbb{R}$ to $e(r) = e^{2\pi i r}$ in S^1, regarded as the unit circle in the complex plane. Since $\mathbb{R}$ is contractible, the pullback $e^*\phi$ is a trivial bundle, and so there is a commutative diagram

$$\begin{array}{ccc} F \times \mathbb{R} & \xrightarrow{E} & X \\ {\scriptstyle pr_2}\downarrow & & \downarrow{\scriptstyle \phi} \\ \mathbb{R} & \xrightarrow{e} & S^1 \end{array}$$

where $F = \phi^{-1}(1)$ is the fibre of ϕ, pr_2 is projection onto the second factor, and E is a covering map. The translation $r \mapsto r+1$ of $\mathbb{R}$ determines a homeomorphism h of F such that $H : (v,r) \mapsto (h(v), r+1)$ generates the covering group of E. The map h is called the characteristic map of the bundle, and determines it up to isomorphism. Although h is only defined up to isotopy, the induced map on homology is well defined, and corresponds to multiplication by t on the homology of the infinite cyclic cover $F \times \mathbb{R}$ of X, considered as a Λ-module. The fibre F is a compact orientable (n+1)-manifold with μ boundary n-spheres, and is connected since the map $\pi_1(\phi) : G \to \mathbb{Z}$ is onto. (For more details on the construction of the characteristic homeomorphism see [133; page 67].)

<u>Theorem 4</u> <u>Let</u> L <u>be a fibred</u> μ-<u>component</u> n-<u>link</u>

(i) <u>If</u> $n = 1$ <u>then</u> $\Delta_{red}(L) \neq 0$.

(ii) <u>If</u> $n > 1$ <u>then</u> $\mu = 1$. <u>In other words</u>, L <u>is an</u> n-<u>knot</u>.

<u>Proof</u> As an abelian group $H_1(X^\tau; \mathbb{Z})$ is finitely generated, since it is isomorphic to $H_1(F; \mathbb{Z})$. Therefore it must be a torsion Λ-module and so, if $n = 1$, $\Delta_{red}(L) \neq 0$.

If $n > 1$ then Stallings' Theorem and Theorem V.1 imply that $E_{\mu-1}(G) = 0$ and so $H_1(X'; \mathbb{Z})$ has rank $\mu - 1$ as a Λ_μ-module. Since the boundary maps in the equivariant chain complex for X^τ may be obtained from those for X' by the change of coefficients $\pi : \Lambda_\mu \to \Lambda$, the rank of $H_1(X^\tau; \mathbb{Z})$ as a Λ-module is at least $\mu - 1$. Hence it can only be finitely generated as an abelian group if $\mu = 1$. //

In view of this theorem we shall continue to concentrate on the case $n = 1$. (For a survey of results on higher dimensional fibred knots, see Section 5 of [98]). The linking number of a fibred 2-component link is unrestricted, as the Whitehead link ($\ell = 0$) and the $(2, 2\ell)$-torus link ($\ell \neq 0$) are fibred. A fibred 2-component link has linking number 0 if and only if the boundary of the fibre is a union of longitudes, for the linking number is the image of either longitude under the total linking number homomorphism when $\mu = 2$. Goldsmith has observed that the (3-component) Borromean rings are fibred [55].

Since the action of h determines the Λ_1-module structure on $H_1(X;\Lambda_1) = H_1(F; \mathbb{Z})$, the characteristic polynomial of $h_1 = H_1(h; \mathbb{Z})$ generates $E_0(H_1(X;\Lambda_1))$ while the minimal polynomial of h_1 generates $\text{Ann}\, H_1(X;\Lambda_1)$. Therefore if $\mu = 1$ the characteristic polynomial is $\Delta_1(L)$ and the minimal polynomial is $\lambda_1(L) = \Delta_1(L)/\Delta_2(L)$, while if $\mu > 1$ the characteristic polynomial is $(t-1)\Delta_{red}(L)$ and the minimal polynomial is $(t-1)\Delta_{red}(L)/\Delta_1(H_1(X;\Lambda_1))$ which divides $(t-1)\pi(\Delta_1(L)/\Delta_2(L))$ since $E_1(H_1(X;\Lambda_1)) = \pi(E_2(L))$ is contained in $(\pi(\Delta_2(L)))$.

The most interesting class of (fibred) links are those associated with an isolated singularity of a plane algebraic curve. Let $f(w,z)$ be a polynomial with complex coefficients which vanishes at the origin and suppose that f has at most an isolated critical point there. If $\varepsilon > 0$ is so small that the ball $B_\varepsilon = \{(w,z) \mid |w|^2 + |z|^2 \leq \varepsilon\}$ contains no critical point of f other than the origin, then the pair $(S_\varepsilon^3, S_\varepsilon^3 \cap f^{-1}(0))$ determines a fibred link whose link type is independent of ε. (The fibration is given by $\phi(w,z) = f(w,z)/|f(w,z)|$ for (w,z) in $S_\varepsilon^3 - f^{-1}(0)$ [133; page 5].) We shall call such a link an algebraic link. The geometry is well understood, as the link is an "iterated torus link" and is specified completely by the Puiseaux expansions of the μ irreducible branches of f at the origin [1,112,184]. (There is one component for each irreducible branch. Note that Milnor and others use μ to denote a different invariant of the singularity [133; page 59].) The action of the generator of $\pi_1(S^1)$ on the homology of the fibre $H_1(F; \mathbb{Z})$ via h_1 is called the (local algebraic) monodromy of f at the origin. In the remainder of this section we shall list some of the more striking results on the monodromy. (We remark that most of these results extend to the higher dimensional case of isolated singularities of hypersurfaces in $\mathbb{C}^{n+1}$, although the proofs in some of the references quoted below depend on the geometry of the classical case.)

Since an algebraic link is an iterated torus link, the roots of its reduced Alexander polynomial are roots of unity, and so the monodromy is quasiunipotent, in other words, $(h_1^{m!} - 1)^N = 0$ for sufficiently large integers m and N [15,112,184]. Lê showed that when $\mu = 1$ the monodromy is of finite order, equivalently that the minimal polynomial $\lambda_1(L)$ has distinct roots [112], but this is not true in general. A'Campo has shown that the link of $(w^2 + z^3)(w^3 + z^2)$ has monodromy of infinite order [1].

Durfee showed that the characteristic polynomial is exactly divisible by $(t-1)^{\mu-1}$, in other words that $\nabla(L)(1) \neq 0$ [46]. In particular a 2-component algebraic link has nonzero linking number, while the Borromean rings cannot be algebraic. He showed also that the monodromy is of infinite order if $(t+1)^{\mu}$ divides the characteristic polynomial. This criterion was extended by Sumners and Woods who showed that if ξ is a p^mth root of 1 for some prime p and positive integer m, and if $(t-\xi)^{\mu}$ divides the characteristic polynomial, then the monodromy is again of infinite order [184].

A'Campo showed that the trace of the monodromy is 1, and deduced that the connected sum of algebraic knots is never algebraic [2]. Lê showed that concordant algebraic knots are isotopic [112].

Brieskorn indicated how the complex monodromy $h_1 \otimes \mathbb{C}$ could (in principle) be derived algebraically from the function f [15] Although the complex monodromy does not determine h_1, it does determine the characteristic polynomial of h_1. In another approach which seems more workable, A'Campo related the characteristic polynomial to a Zeta-function [3]. Durfee has shown that there is a basis for $H_1(F;\mathbb{Z})$ for which the Seifert matrix is upper triangular with entries -1 on the diagonal [45].

Finite Cyclic Covers

Below the total linking number cover $q : X^{\tau} \to X$ lie the k-fold cyclic covers $q_k : X_k \to X$ corresponding to the subgroups $\tau^{-1}(k\mathbb{Z})$ of G. The space X_k is just X^{τ} modulo the action of t^k, and there is an infinite cyclic cover $X^{\tau} \to X_k$. Thus the homology of the finite cyclic covering spaces may be determined (up to an extension problem) from that of the total linking number cover via Wang sequences. Conversely Cowsik and Swarup have shown that the integral homology of X^{τ} imbeds in the inverse limit of the homologies of the X_k [35]. (They note that this need not be true of the rational homology. For example, consider any knot, such as 4_1, for which the Alexander polynomial is nontrivial and has a non-cyclotomic factor.) They have thereby simplified and extended arguments of Gordon [58] and Durfee and Kauffman [47] on periodicity in the homology of branched covering spaces of knots.

The finite cyclic covers of the exterior of a link may be completed to branched covers of the sphere. (See [157; page 292]). When $n = 1$ this is a fruitful way of constructing closed 3-manifolds. Branched cyclic covers are also central to the most natural proof of the invariance of the Hermitean signatures of a link under concordance [88, 198].

The Wang sequence for the cover $X^\tau \to X_k$ in low dimensions reduces to

$$\to H_1(X^\tau; \mathbb{Z}) \xrightarrow{t^k-1} H_1(X^\tau; \mathbb{Z}) \to H_1(X_k; \mathbb{Z}) \xrightarrow{\partial} \mathbb{Z} \to 0 .$$

The image of the kth power of a meridian in $H_1(X_k; \mathbb{Z})$ generates a submodule mapped isomorphically onto $\mathbb{Z}$ by ∂, and so $H_1(X_k; \mathbb{Z}) \approx \mathbb{Z} \oplus C_k$, where $C_k = H_1(X^\tau; \mathbb{Z})/(t^k - 1)H_1(X^\tau; \mathbb{Z})$. Let $M_k = X_k \cup \mu D^2 \times S^1$ be the k-fold cyclic branched covering space of a 1-link L. Then Shinohara and Sumners show that the map $H_2(M_k, X_k; \mathbb{Z}) \to H_1(X_k; \mathbb{Z})$ is injective [172]. Using this Sumners computes the Betti numbers of M_k [183]. If L is a knot ($\mu = 1$), then $H_1(M_k; \mathbb{Z}) \approx C_k$, and Weber shows then that C_k is finite, of order $|Res(t^k - 1, \Delta_1(L))| = |\prod_{0 \leq i < k} \Delta_1(L)(\xi^i)|$ where ξ is a primitive k^{th} root of unity, if and only if this number is nonzero [203]. Since $|\Delta_1(L)(1)| = 1$ for a knot, this number may be rewritten as $|Res((t^k - 1/t - 1), \Delta_1(L))| = |\prod_{1 \leq i < k} \Delta_1(L)(\xi^i)|$. Using a presentation for $\pi_1(X_k)$ derived from a Wirtinger presentation for G by a Reidemeister-

Schreier process, Kinoshita shows that in the latter formulation the assertion remains true if $\mu > 1$. [104a]. (Note that he has observed that for the 2-component (2,4)-torus link  the first homology of M_2 is $\mathbb{Z}/4\mathbb{Z}$, and is not a direct summand of that of X_2, which is $\mathbb{Z}^2 \oplus \mathbb{Z}/2\mathbb{Z}$. See [172]).

A late addition. Sakuma has given a neat calculation of the order of the homology of a cyclic cover of S^3, branched over a link [209]. By the Cartan-Leray spectral sequence for $r: X' \to X^\tau$, the sequence

$$0 \to R(= r_* H_1(X')) \to H_1(X^\tau) \to \mathbb{Z}^{\mu-1} \; (= Tor_1^{\Lambda_\mu}(\Lambda_1, \mathbb{Z})) \to 0$$

is exact. Sakuma observes $H_1(M_k) = H_1(X^\tau)/\nu H_1(X^\tau)$, where $\nu = (t^k - 1/t - 1)$, and hence that there is an exact sequence

$$0 \to R/\nu R \to H_1(M_k) \to (\mathbb{Z}/k\mathbb{Z})^{\mu-1} \to 0.$$

If $\nabla(L) \neq 0$, then $H_1(X^\tau)$ has no pseudozero submodule and (hence) R has a square presentation matrix. By Lemma III.6 and page 102, $\Delta_0(R) = \nabla(L)$. By the argument of Weber [203], the order of $R/\nu R$ is $|Res(\nabla(L),\nu)|$ and so that of $H_1(M_k)$ is $|Res(\Delta_{red},\nu)|$. (Note that $|Res(t-1,\nu)| = k$). Clearly $H_1(M_k)$ is finite if and only if this number is nonzero. Sakuma's calculation of $|\varepsilon(\nabla)|$ is equally neat. The Wang sequence for $q: X^\tau \to X$ leads to

$$H_2(X) \to H_1(X^\tau) \to R \to 0$$

(as $R = (t-1)H_1(X^\tau)$, by an earlier sequence) and hence to

$$H_2(X) \xrightarrow{U} H_1(X^\tau)/(t-1)H_1(X^\tau) \; (= \ker \partial : H_1(X) \to H_0(X^\tau)) \to R/(t-1) \to 0$$

where the matrix of U with respect to suitable bases is a minor of the matrix N of page 105, so $|\varepsilon(\nabla)| = |\Delta_0(R/(t-1)R)| = |E_1(N)|$.

CHAPTER IX LOCALIZING THE BLANCHFIELD PAIRING

A (2q-1)-knot is determined up to concordance by the equivalence class of its Blanchfield pairing in a certain Witt group, if $q \geqslant 2$ [94,97,118,180]. (Kervaire had shown earlier that all even dimensional knots are null concordant [96]). This pairing is also defined and of considerable interest in the classical case $q = 1$, although it is no longer a complete invariant of concordance [24]. In this chapter we shall propose similar invariants for classical links. The Blanchfield pairing on the quotient of the Λ_μ-torsion of the Alexander module of a 1-link L by its maximal pseudozero submodule, after localization with respect to S, the multiplicative system in Λ_μ generated by the elements $t_1-1,\ldots t_\mu-1$, is a primitive Hermitean pairing into $Q(t_1,\ldots t_\mu)/\Lambda_{\mu S}$. We shall show that the class of this pairing in the Witt group of such pairings depends only on the concordance class of L. If local knotting is factored out, by passing to the coarser equivalence relation of weak concordance, generated by concordance and isotopy, the appropriate invariant is obtained by localizing further with respect to the multiplicative system generated by all the nonzero polynomials in one variable. (This more thorough going localization was motivated by Rolfsen's result that the "Alexander invariant" localized at Σ is invariant under isotopy [155]).

These invariants may be computed for a boundary link from a Seifert matrix for the link. The invariant of concordance specializes to the knot concordance invariant of [94] for 1-component links; the weak concordance invariant is clearly trivial for knots. For 2-component links, the coefficient ring Λ_Σ is a principal ideal domain, so a primitive Λ_Σ-torsion pairing is perfect, and the Witt group of linking pairings over Λ_Σ can be expressed as a direct sum of Witt groups of Hermitean forms over involuted fields. We shall show that the image of the set of 2-component boundary links in this Witt group is a subgroup which is not finitely generated. In the next section we refer briefly to several definitions of signatures for links, in order to raise the question as to how they are related to the Witt classes introduced here. We conclude this chapter with an appendix in which we use a surgery description of a knot in order to describe its Blanchfield pairing.

Linking Pairings on Torsion Modules

Let R be a commutative noetherian domain with an involution $-:R \to R$, and with field of fractions $F = R_o$. We shall use the overbar also to denote the involution induced on F and on the R-module F/R. Let $\varepsilon = \pm 1$. (There will be no risk of confusion with the augmentation homomorphism in this chapter).

Definition A map $c:N_1 \times N_2 \to N_3$ of R-modules is a sesquilinear pairing if it is $\mathbb{Z}$-bilinear, R-linear in the first variable and R-antilinear in the second variable (that is, $c(rn_1,n_2) = rc(n_1n_2) = c(n_1,\bar{r}n_2)$ for all n_1 in N_1, n_2 in N_2 and r in R). The pairing c is primitive on the left (respectively, right) if for each nonzero n_1 in N_1 (respectively, n_2 in N_2) there is some n_2 in N_2 (respectively, n_1 in N_1) such that $c(n_1,n_2) \neq 0$.

An ε-Hermitean pairing on a torsion module M is a sequilinear pairing $b:M \times M \to F/R$ such that $b(m,n) = \varepsilon\, \overline{b(n,m)}$ for all m and n in M. The adjoint of b is the R-antilinear map $Ad(b):M \to Hom_R(M,F/R)$ defined by $Ad(b)(n)(m) = b(m,n)$ for all m and n in M. The pairing b is primitive if Ad(b) is injective and perfect if Ad(b) is bijective.

If N is a finitely generated R-module and P is the maximal pseudozero submodule of N we shall let $\hat{t}N = tN/P$. (Of course if R is a P.I.D. then $tN = \hat{t}N$ for any R-module N). As remarked in Chapter III, if a torsion module M supports a primitive bilinear (or sequilinear) pairing, it can have no nontrivial pseudozero submodule, so $M = \hat{t}M$.

Lemma 1 If R is a principal ideal domain, then a primitive ε-Hermitean pairing on a finitely generated torsion module M is perfect.

proof By the structure theorem for finitely generated modules over P.I.D.s, M is a direct sum of cyclic modules, and so $\mathrm{Hom}_R(M,F/R)$ is (noncanonically) isomorphic to M. As M has finite length, any injective (antilinear) endomorphism is easily seen to be bijective. //

Definition An ε-linking pairing over R is a finitely generated torsion module M with a primitive ε-Hermitean pairing b. The sum of two such pairings (M,b) and (M',b') is the pairing $(M,b) \oplus (M',b')$ with underlying module $M \oplus M'$ and with map sending $(m_1 \oplus m_1', m_2 \oplus m_2')$ to $b(m_1,m_2) + b'(m_1',m_2')$. A pairing is neutral if M contains a submodule N such that $N = N^{\perp} = \{m \text{ in } M | b(n,m) = 0 \text{ for all } n \text{ in } N\}$. Two pairings (M,b) and (M',b') are Witt equivalent if there are neutral pairings (N,c) and (N',c') such that $(M,b) \oplus (N,C) \approx (M',b') \oplus (N',c')$.

Proposition-Definition The set of Witt-equivalence classes of ε-linking pairings, with addition defined by sum of representative pairings, and with (M,-b) representing the inverse of the class of (M,b), is an abelian group, denoted $W_\varepsilon(F,R,-)$. //

The Witt groups of greatest interest to the algebraist are based on perfect pairings with an additional quadratic structure. If 2 is invertible in R, any perfect pairing can be endowed uniquely with such a quadratic structure, and so the relative Witt group $W_0^{\varepsilon}(F/R)$ of Pardon [148] then embeds in $W_\varepsilon(F,R,-)$ (since in any case the definition of neutral is the same for primitive and perfect pairings). All these distinctions vanish when R is a P.I.D. containing $\frac{1}{2}$, as in the ring $\Lambda_{2\Sigma}$ discussed below in connection with 2-component links.

For the remainder of this section we shall assume that R is factorial.

lemma 2 (Blanchfield) If $c: N_1 \times N_2 \to F/R$ is a sesquilinear pairing which is primitive on both sides, then $(\Delta_o(N_1)) = (\overline{\Delta_o(N_2)})$.

proof This is an immediate consequence of Theorem 4.5 of Blanchfield [11]. Alternatively, it may be proven by localizing with respect to the multiplicative system $R-(\mathfrak{p}\cup\bar{\mathfrak{p}})$, for each height 1 prime $\mathfrak{p}$ of R. Compare our proof of Theorem VII.1. //

lemma 3 If (N,c) is a neutral ε-linking pairing over R, then $(\Delta_o(N)) = (f\bar{f})$ for some nonzero f in R.

proof By assumption there is a submodule $P \subset N$ such that $P = P^{\perp}$. The pairing c induces a sesquilinear pairing of P and N/P into F/R which is primitive on both sides. By the preceding lemma $(\Delta_o(P)) = (\overline{\Delta_o(N/P)})$. Therefore by lemma III.6 $(\Delta_o(N)) = (\Delta_o(P))(\Delta_o(N/P)) = (\Delta_o(P))(\overline{\Delta_o(P)})$. Let f be any generator of $(\Delta_o(P))$.//

Theorem 4 If (M,b) and (M',b') are Witt-equivalent ε-linking pairings over R then there are nonzero elements f and f' in R such that $(f\bar{f}\Delta_o(M)) = (f'\bar{f}'\Delta_o(M'))$. Therefore the map sending (M,b) to the class of $\Delta_o(M)$ modulo $\{ug\bar{g} \mid u \text{ in } R^*,\ g \text{ in } F^*\}$ induces a homomorphism from $W_\varepsilon(F,R,-)$ to $\hat{H}^o(\mathbb{Z}/2\mathbb{Z}: F^*/R^*) =$

$$\{vf \mid v \text{ in } R^*,\ f = \bar{f} \text{ in } F^*\}/\{ug\bar{g} \mid u \text{ in } R^*,\ g \text{ in } F^*\}.$$

proof By assumption there are neutral pairings (N,c) and (N',c') such that $(M,b) \oplus (N,c) \approx (M',b') \oplus (N',c')$. The first assertion now follows from the lemmas. The rest is clear. //

Definition The Alexander class of (the Witt class of) an ε-linking pairing (M,b) is the image $\delta(M)$ of $\Delta_o(M)$ in $\hat{H}^o(\mathbb{Z}/2\mathbb{Z}; F^*/R^*)$.

We shall let K_μ (or K) denote the field of fractions $Q(t_1,\dots t_\mu)$ of Λ_μ. The multiplicative systems in Λ_μ generated by $\{t_1-1,\dots t_\mu-1\}$ and by $\bigcup_{1<i<\mu} \mathbb{Z}[t_i,t_i^{-1}]-\{0\}$ are denoted by S and Σ respectively.

Blanchfield Duality

In this section we shall give an explicit geometric description of a version of equivarent duality, and use it to define our invariant. Blanchfield showed that if $p:\tilde{X}\to X$ is a connected cover of a compact oriented n-manifold X, with covering group $\mathbb{Z}^\mu$, then there is a sesquilinear pairing $V:\hat{t}H_p(\tilde{X})\times\hat{t}H_{n-1-p}(\tilde{X},\partial\tilde{X})\to K/\Lambda$ which is primitive on both sides [11]. (Here and below in this chapter we shall let $H_*(Y)$ denote the integral homology of the space Y). If x is a p-cycle of $\tilde{X}$ such that $\alpha x=\partial u$ for some $(p+1)$-chain u and some nonzero α in Λ, and y is a relative $(n-1-p)$-cycle such that $\beta y=\partial v$ for some $(n-p)$-chain v and some nonzero β in Λ, then $V(x,y)=\alpha^{-1}S(u,y)=\bar{\beta}^{-1}S(x,v)$ where $S(c,d)=\sum_{\lambda \text{ in } \mathbb{Z}^\mu} I(c,\lambda d)\cdot\lambda$ and I is ordinary intersection of chains. From this definition, it is not hard to see that if $n=2p+1$, then V induces a $(-1)^{p+1}$-Hermitean pairing $[\ ,\ ]$ on $\hat{t}H_p(\tilde{X})$ by $[x,y]=V(x,j_*y)$ (where j is the inclusion $(\tilde{X},\emptyset)\to(\tilde{X},\partial X)$) which however may not be primitive. He showed also that intersection numbers give a nonsingular pairing of the "Betti" modules $H_p(\tilde{X})/tH_p(\tilde{X})$ and $H_{n-p}(\tilde{X},\partial\tilde{X})\,/\,tH_{n-p}(\tilde{X},\partial\tilde{X})$ into Λ.

Now let L be a μ-component 1-link with complement X, and let $p:X'\to X$ be the maximal abelian cover of X. Then the meridians of L determine an isomorphism of the covering group with $\mathbb{Z}^\mu$. Then the Blanchfield linking pairing is a map $\hat{t}H_1(X')\times\hat{t}H_1(X',\partial X')\to K/\Lambda$. In the case of knots $H_1(X')$ is a Λ_1-torsion module and the natural map $H_1(X')\to H_1(X',\partial X')$ is an isomorphism. In general however, the two outer maps of the following part

$$H_1(\partial X') \to H_1(X') \to H_1(X',\partial X') \to H_0(\partial X')$$

of the long exact sequence of the pair $(X',\partial X')$ are non-zero. Nevertheless $H_1(\partial X')$ and $H_0(\partial X')$ vanish after localizing with respect to S (and, a fortiori, with respect to Σ), since both are quotients of $\bigoplus_{i=1}^{\mu} (\Lambda/t_i-1))$. Thus the localized Blanchfield pairings

$$[-,-]_S : \hat{t}H_1(X')_S \times \hat{t}H_1(X')_S \to K/\Lambda_S$$

and

$$[-,-]_\Sigma : \hat{t}H_1(X')_\Sigma \times \hat{t}H_1(X')_\Sigma \to K/\Lambda_\Sigma$$

are primitive, (+1)-Hermitian pairings. (Notice that the Λ_S-torsion of M_S is the localization of the Λ-torsion of M, so the notation tM_S is unambiguous. Note also that $A(L)_S = H_1(X')_S \oplus \Lambda_S$ and $\hat{t}H_1(X')_S = \hat{t}A(L)_S$).

Definition For L a μ-component link, $B_S(L)$ (respectively. $B_\Sigma(L)$) is the class of $(\hat{t}H_1(X')_S, [-,-]_S)$ (respectively, $(\hat{t}H_1(X')_\Sigma, [-,-]_\Sigma)$) in $W_{+1}(K_\mu,\Lambda_{\mu S},-)$ (respectly, $W_{+1}(K_\mu,\Lambda_{\mu\Sigma},-)$).

For knots multiplication by $1 - t$ induces an automorphism of $H_1(X')$ [96] and so essentially no information is lost on localization with respect to S in that case. Localization with respect to Σ annihilates $H_1(X')$ for X' the infinite cyclic cover of a knot complement; but that is all that is lost; Rolfsen showed that any PL isotopy of a link could be achieved by introducing or suppressing local knots, and that hence localizing with respect to Σ the homology of the maximal abelian cover of a link complement gave an isotopy invariant [155]. It is easy to see that the Σ-localised Blanchfield pairing is also invariant under such local isotopy, and hence under isotopy.

Computation from a Seifert Matrix

The computation of the Blanchfield pairing from a knot in terms of a Seifert matrix, as done by Kearton [93], may also be carried through for boundary links. Let $L : \mu S^1 \to S^3$ be a boundary link, and let U_j, $1 \leq j \leq \mu$, be μ disjoint orientable surfaces spanning L. Then orientations for the surfaces U_j compatible with those of the L_j are determined uniquely by the convention inward normal last (equivalently, by insisting that for each j the orientation class $[U_j, \partial U_j]$ in $H_2(U_j, \partial U_j)$ map to $[L_j]$ in $H_1(\partial U_j)$ in the long exact sequence of the pair $(U_j, \partial U_j)$). Let $Y = X - \bigcup_{i=1}^{\mu} W_j$ where the W_j are disjoint open regular neighbourhoods of U_j in the link complement X, $1 \leq j \leq \mu$. There are two natural embeddings of each U_j in Y; call the one for which U_j is compatibly oriented with ∂Y i_{j+} and the other i_{j-}. Then the module $H_1(X')$ is contained in the following segment of a Mayer-Vietoris sequence:

$$H_1(U) \otimes_{\mathbb{Z}} \Lambda \xrightarrow{d_1} H_1(Y) \otimes_{\mathbb{Z}} \Lambda \to H_1(X') \to H_o(U) \otimes_{\mathbb{Z}} \Lambda \xrightarrow{d_1} H_o(Y) \otimes_{\mathbb{Z}} \Lambda \to \mathbb{Z} \to 0$$

where $U = \bigcup_{j=1}^{\mu} U_j$ and $d_* \mid H_*(U_j) \otimes \Lambda = (i_{j+})_* \otimes t_j - (i_{j-})_* \otimes 1$.

<u>Lemma 5</u> $\hat{t}H_1(X') = tH_1(X) = \text{Coker } d_1$.

<u>Proof</u> Clearly $tH_1(X') \subseteq \text{Coker } d_1$, since $H_0(U) \otimes \Lambda$ is free. On localizing the above sequence with respect to K, we obtain the exact sequence

$$0 \to (\text{Coker } d_1) \otimes_\Lambda K \to H_1(X') \otimes_\Lambda K \to K^\mu \to K \to 0.$$

On the other hand a similar localization of the Crowell sequence relating $H_1(X')$ to the Alexander module gives

$$0 \to H_1(X') \otimes_\Lambda K \to A(L) \otimes_\Lambda K \to K \to 0$$

and $A(L) \otimes_\Lambda K = K^\mu$ for a μ-component boundary link (since then $E_{\mu-1}(L) = 0$ and $E_\mu(L) \neq 0$). Hence $(\text{Coker } d_1) \otimes_\Lambda K = 0$ and so Coker d_1 is a torsion module.

Let $\{\alpha_{jm} \mid 1 \leq m \leq m(j)\}$ be a basis for $H_1(U_j)$, $1 \leq j \leq \mu$. By Alexander duality, the ordinary linking pairing ℓ in S^3 [177; page 361] establishes a duality between $H_1(U)$ and $H_1(Y) = H_1(S^3-U)$; let $\{\hat{\alpha}_{jm} \mid 1 \leq m \leq m(j),\ 1 \leq j \leq \mu\}$ be the dual basis, so that $\ell(\alpha_{jm}, \hat{\alpha}_{kn}) = \delta_{jk}\delta_{mn}$. Let A be the matrix of

$$(i_+)_* = \bigoplus_{j=1}^{\mu} (i_{j+})_* : H_1(U) \to H_1(Y)$$

with respect to these bases; then d_1 is represented by the matrix $\Delta = \Theta A - A^{tr}$ where Θ is the diagonal matrix $\mathrm{diag}\,(t_1 I_{m(1)}, \ldots t_\mu I_{m(\mu)})$, and $\det \Delta \neq 0$ since it generates $E_0(\mathrm{Coker}\, d_1)$. Then there is an exact sequence

$$0 \to \Lambda^M \to \Lambda^M \xrightarrow{\phi} tH_1(X') \to 0$$

(where $M = \sum_{j=1}^{\mu} m(j)$). Therefore by Theorem III.10 $tH_1(X')$ has no nontrivial pseudozero submodule, and so $\hat{t}H_1(X') = t(H_1(X')$. //

Using the same symbols to denote 1-chains in U, Y representing the classes α_{jm}, $\hat{\alpha}_{kn}$ respectively, consider the 2-chain $[-1,1] \times \alpha_{jm}$ in X'. Then $\partial([-1,1] \times \alpha_{jm}) = t_j i_{j+}(\alpha_{jm}) - i_{j-}(\alpha_{jm})$ and so $\partial\left([-1,1] \times \Sigma\, q_{jm}\alpha_{jm}\right)$ represents $\Sigma\, (\Delta q)_{kn}\hat{\alpha}_{kn}$. Therefore, if $\delta = \det \Delta$, $\delta \cdot \Sigma\, r_{kn}\hat{\alpha}_{kn}$ is the class of $\partial\left([-1,1] \times \Sigma\, (\Delta^{-1}\delta\cdot r)_{jm}\alpha_{jm}\right)$, where the matrix $\Delta^{-1}\cdot\delta$ has coefficients in Λ, and so the Blanchfield pairing on $tH_1(X')$ is given by

$$[\phi(r), \phi(s)] = V(\phi(r), j_*\phi(s)) = \frac{1}{\delta}\, S\left([-1,1] \times \Sigma\, (\Delta^{-1}\delta r)_{jm}\alpha_{jm}, \Sigma\, s_{kn}\hat{\alpha}_{kn}\right)$$

$$= \frac{1}{\delta}\, \Sigma\, \bar{s}_{kn}\, \Sigma\, (\Delta^{-1}\delta r)_{jm} S([-1,1] \times \alpha_{jm}, \hat{\alpha}_{kn})$$

$$= \frac{1}{\delta}\, \Sigma\, \bar{s}_{kn}\, \Sigma\, (\Delta^{-1}\delta r)_{jm}(1-t_j)\delta_{jk}\delta_{mn} = \bar{s}^{tr}(I-\Theta)\cdot\Delta^{-1} r \bmod \Lambda$$

(where r and s are column vectors in Λ^M). Thus for boundary links the Hermitian pairing on $\hat{t}H_1(X')$ is primitive, and in fact perfect, even before localization. (Notice that $\det\Delta$ represents the Alexander class here).

A similar construction works for homology boundary links, for in [73] it is shown that splitting along singular Seifert surfaces $V = \bigcup_{1 \leq i \leq \mu} V_i$ for such a link leads to a short free resolution

$$(H_1(V)/H_1(\partial V)) \otimes \Lambda \to (H_1(X-V)/H_1(\partial X-V)) \otimes \Lambda \to tH_1(X')/B \to 0$$

where B is the submodule generated by the image of the longitudes. By Theorem VI.2, the module $tH_1(X')/B$ is $\hat{t}H_1(X')$. It is however better to localize and thus have an invariant applicable to all links, for the set of (homology) boundary links is not closed under concordance, as is shown by the examples of figures V.1 and VI.1.

Invariance Under Concordance

The above construction of the Blanchfield pairing for boundary links in terms of Seifert surfaces can be used to show that $B_S(L) = B_S(\tilde{L})$ if there is a concordance $\mathscr{L}$ from L to $\tilde{L}$ with group $\mathscr{G} = \pi_1(S^3 \times I - \text{im } \mathscr{L})$ such that $\mathscr{G}/\mathscr{G}_\omega$ is freely generated by a set of meridians, by imitating Levine's proof that when $\mu = 1$, $B_S(k)$ depends only on the concordance class of the knot k [118]. As it is not true in general that every concordance between boundary links is a "boundary concordance" in this sense, let alone that every link concordant to a boundary link is a boundary link, an argument which does not rely on Seifert surfaces is to be preferred.

Theorem 6 Let L_0 and L_1 be concordant μ-component links. Then $B_S(L_0) = B_S(L_1)$.

Proof Let $\mathscr{L} : \mu S^1 \times I \to S^3 \times I$ be a concordance from $L_0 = \mathscr{L} | \mu S^1 \times \{0\}$ to $L = \mathscr{L} | \mu S^1 \times \{1\}$. Let $N(\mathscr{L})$ be an open regular neighbourhood for the image of $\mathscr{L}$ on $S^3 \times I$, and let $Z = S^3 \times I - N(\mathscr{L})$. Then

$\partial Z = X_0 \cup \mu(S^1 \times S^1 \times 1) \cup X_1 \approx X_0 \underset{\mu S^1 \times S^1}{\cup} X_1$ where $X_i = Z \cap S^3 \times \{i\}$ is complement of L_i (for $i=0,1$) and where the jth boundary component of X_0 is identified with the jth boundary component of X_1 via an orientation reversing map. The inclusion $X_0 \to Z$, $X_1 \to Z$ each induce isomorphisms on homology. On localizing the Mayer-Vietoris sequence of the triple $(\partial Z', X_0', X_1')$ with respect to S, it follows that $H_1(\partial Z)_S = H_1(X_0)_S \oplus H_1(X_1)_S$. Clearly the Blanchfield pairing on $\hat{t}H_1(\partial Z)_S$ is the direct sum of the Blanchfield pairing on $tH_1(X_0)_S$ with the negative of the Blanchfield pairing on $\hat{t}H_1(X_1)_S$. Thus to show that $B_S(L_0) = B_S(L_1)$ it will suffice to show that $\hat{t}H_1(\partial Z')_S$ contains a submodule equal to its own annihilator with respect to the pairing.

A. Kawauchi showed in [90; lemma 2.1] that $H_2(Z', X_0')$ is a torsion Λ-module, and hence that the image of $H_2(Z')$ in $H_2(Z', \partial Z')$ is contained in the torsion submodule. Therefore the sequence $tH_2(Z', \partial Z') \xrightarrow{\partial} tH_1(\partial Z') \to tH_1(Z')$ is exact. Let $P = \partial[tH_2(Z', \partial Z')]$. Then the image of P_S in $\hat{t}H_1(\partial Z')_S$ is such a submodule.

For let Q, R be relative 2-cycles on $(Z', \partial Z')$ representing torsion classes in $H_2(Z', \partial Z')$ and let q, r be the boundaries of Q, R respectively, 1-cycles on $\partial Z'$ representing classes in P. Then $\alpha r = \partial s$ for some nonzero α in Λ and for some 2-chain s on $\partial Z'$. Then $V_{\partial Z'}(q,r) = \frac{1}{\bar{\alpha}} \sum_{\gamma} I_{\partial Z'}(q, \gamma s)\gamma = \frac{1}{\bar{\alpha}} \sum_{\gamma} I_{Z'}(Q, \gamma \hat{s}) = V_{Z'}(Q, \hat{r})$ (where $\hat{r}$, $\hat{s}$ denote r, s considered as chains on Z') = 0 in K/Λ since $\hat{r}$ bounds R in X'. Thus $P \subset P^{\perp}$. Now let w be a 1-cycle on $\partial Z'$ representing a torsion class in $tH_1(\partial Z')$ (so that $\beta w = \partial W$ for some nonzero β in Λ and some 2-chain W on $\partial Z'$) and suppose $V_{\partial Z'}(q,w) = 0$ for every 1-cycle q as above, representing a class in P. Then $V_{Z'}(Q, \hat{w}) = V_{\partial Z'}(q,w)$, and hence by the primitivity of the Blanchfield pairing for $(Z', \partial Z')$, $\hat{w}$

bounds in Z'. Hence the class represented by w is in the image of d, and so $P^{\perp} = P$. It follows immediately that $P_S^{\perp} = P_S$, and so the theorem is proven. //

Corollary <u>If</u> L <u>is a null concordant link,</u> $B_S(L) = 0$.//

The analogous results for $B_\Sigma(L)$ are also immediate consequences of this theorem, namely $B_\Sigma(L_0) = B_\Sigma(L_1)$ if L_0 is weakly concordant to L_1, and in particular $B_\Sigma(L) = 0$ if L is weakly concordant to the trivial μ-component link.

Corollary <u>Let</u> $\delta_S(L)$ <u>denote the least principal ideal in the</u> U.F.D. Λ_S <u>containing</u> $E_0(t(A(L))_S = E_0(tH_1(X(L)')_S)$. <u>Then</u> $\delta_S(L) = \overline{\delta_S(L)}$ <u>and if</u> L_0 <u>and</u> L_1 <u>are concordant, then there are nonzero</u> f_0, f_1 <u>in</u> Λ_S <u>such that</u> $f_0 \cdot \bar{f}_0 \cdot \delta_S(L_0) = f_1 \cdot \bar{f}_1 \cdot \delta_S(L_1)$. <u>(Similarly for localization with respect to</u> Σ).

Proof These assertions are consequences of the discussion above of the determinant of a linking pairing, and of the result $B_S(L_0) = B_S(L_1)$. //

The ideal $\delta_S(L)$ is generated by the image of the first nonzero Alexander polynomial of the link. Kawauchi has proven the stronger result that if L_0 and L_1 are concordant links, then there are nonzero g_0 and g_1 in Λ with $\varepsilon(g_0) = \varepsilon(g_1) = 1$ and such that $g_0 \cdot \bar{g}_0 \cdot \Delta_\alpha(L_0) = g_1 \cdot \bar{g}_1 \cdot \Delta_\alpha(L_1)$ (where $\alpha = \alpha(L_0) = \alpha(L_1)$) [90].

Additivity

Let L_-, L_+ be μ-component links. After an ambient isotopy of each link, it may be supposed that $\mathrm{im}L_- \subset D^3_-$, $\mathrm{im}L_+ \subset D^3_+$ (where $S^3 = D^3_- \cup_{S^2} D^3_+$) and that for each i, $1 \leq i \leq \mu$, the ith component of $\mathrm{im}L_-$ meets $\mathrm{im}L_+$ only in an arc contained in the ith component of $\mathrm{im}L_+$, which receives opposing orientations from L_- and L_+. Then the closure of

$$\mathrm{im}L_- \cup \mathrm{im}L_+ \;\; - \mathrm{im}L_- \cap \mathrm{im}L_+$$

is the image of a compatibly oriented link. If $\mu = 1$, the ambient isotopy type of this link is well defined by the ambient isotopy types of L_- and L_+. Let C_1^{μ} denote the set of concordance classes of μ-component links. Then this connected sum induces an addition on $C_1 = C_1^1$, the set of concordance classes of knots, so that C_1 becomes an abelian group, and the Blanchfield pairing gives rise to a homomorphism from C_1 to $W_+(Q(t), \mathbb{Z}[t,t^{-1}], -)$ [94, 118]. It is not true in general that the connected sum of two links is well defined, even modulo concordance, Already with $L_- = -L_+ =$  (the two component abelian link) connected sums may be formed in at least two non-concordant ways. The link of Figure 1a is trivial, whereas the link of Figure 1b has first Alexander ideal nonzero, and so these two links are not even I-equivalent .

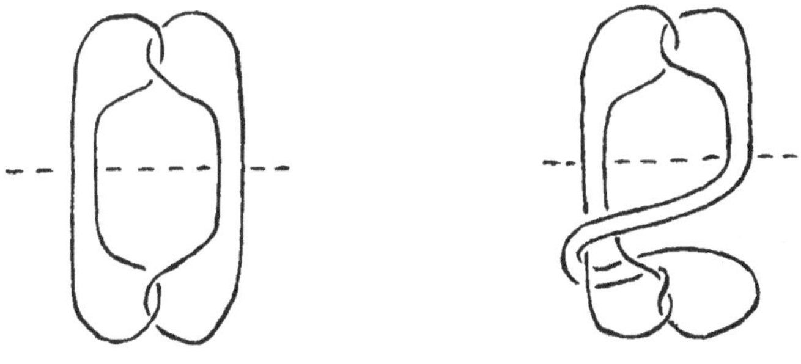

(a) (b)

Figure 1

Furthermore, the Alexander module of the link of Figure 2, which is a connected sum of the abelian link with itself, is $\Lambda/(1+t_1t_2)$ which when

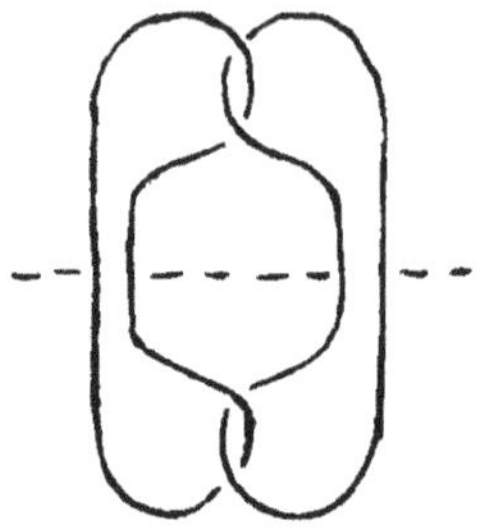

Figure 2

localised with respect to Σ has length 1 as a Λ_Σ-module, and so can not support any pairing whose Witt class is divisible by 2, as would be expected if the Blanchfield pairing were always 'additive', for modules admitting such pairings necessarily have even length. Nevertheless some additivity results may be obtained by restricting the classes of links considered. Let $L_1 \# L_2$ denote any link formed in the above fashion from two links $L_1, L_2 : \mu S^1 \to S^3$.

In the first place, if $\mathscr{L} : \mu S^1 \times I \to S^3 \times I$ is a concordance then $\mathscr{L} \mid \mu * \times I$ embeds μ disjoint arcs and so by general position and by the isotopy extension theorem [159; page 56] it may be assumed that $\mathrm{im}\mathscr{L}$ is contained in $D^3_- \times I$ and meets $S^2 \times I$ in (μ arcs) $\times$ I. Hence if L_1 and L_2 are concordant to L_1' and L_2' respectively, then any link of the form $L_1 \# L_2$ is concordant to some link of the form $L_1' \# L_2'$. In particular the set of concordance classes containing split links forms an abelian group, isomorphic to $(C_1)^{\mu}$, which acts on C_1^{μ}. In fact C_1^{μ} is $(C_1)^{\mu}$-equivariantly isomorphic to $(C_1)^{\mu} \times \bar{C}_1^{\mu}$, where $\bar{C}_1^{\mu}$ is the set of weak concordance classes of μ-component links. (The map $\bar{C}_1^{\mu} \to C_1^{\mu}$ sends the weak concordance class of L to the concordance class of $L \# \left(\coprod_{i=1}^{\mu} (-L_i) \right)$ where $\coprod_{i=1}^{\mu} (-L_i)$ is the split link whose i^{th} component is the reflected inverse of the ith component of L.)

Secondly, if L_1 and L_2 are both boundary links, and if L_1 (respectively L_2) bounds a system of disjoint Seifert surfaces U_{1i} in D^3_- (respectively U_{2i} in D^3_+), then $L_1 \# L_2$ is a boundary link with a system of Seifert surfaces given by $U_{1i} \# U_{2i}$. It is then clear from the interpretation of the Blanchfield pairing in terms of Seifert surfaces that $B_S(L_1 \# L_2) = B_S(L_1) + B_S(L_2)$. (Not every connected sum of boundary links is a boundary link though, for the ribbon link of Figure 3, which is not even an homology boundary link, is the connected sum of two copies of the trivial link.

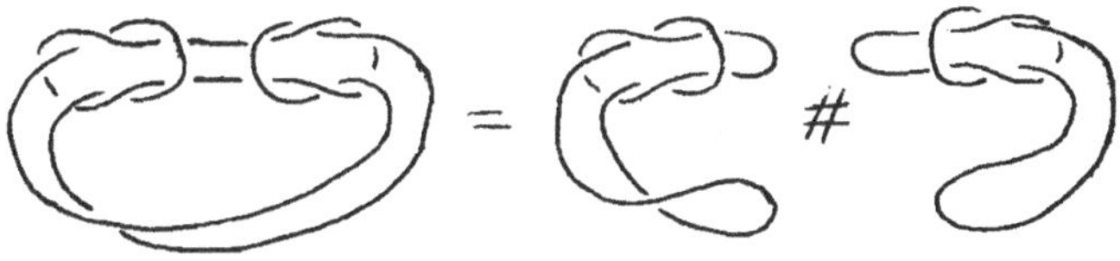

Figure 3.

For this link, and for the trivial link, the Alexander module is torsion free, and so the Blanchfield pairings still add.) This shows that the image of the class of boundary links in $W_{+1}(K, \Lambda_S, -)$ is a subgroup (Since $B_S(-L) = -B_S(L)$ it is clearly closed under taking inverses).

Thirdly, if L_1 and L_2 are μ-component links with Alexander nullity $\alpha(L_1) = \alpha(L_2) = \mu$, then $\alpha(L_1 \# L_2) = \mu$ also. (More precisely, if a link $\tilde{L}$ is obtained from a link L by one saddle point amalgamation [49] then an examination of Jacobian matrices shows that $\alpha(\tilde{L}) \geqslant \alpha(L) - 1$. Hence $\alpha(L_1 \# L_2) \geqslant \alpha(L_1) + \alpha(L_2) - \mu$ if L_1 and L_2 are both μ-component links). This is of some interest as the set of all such links contains all (homology) boundary links, and is closed under concordance by Theorem V.2. Is every link L with $\alpha(L) = \mu$ concordant to a boundary link? If L_1 is a boundary link in D^3_- (as above) and L_2 is a link in D^3_+ with $\alpha(L_2) = \mu$ which meets L_1 in μ disjoint arcs in $S^2 = D^3_- \cap D^3_+$, then it can be shown that $B_S(L_1 \# L_2) = B_S(L_1) + B_S(L_2)$. Is it generally true that the Blanchfield pairing is additive for links L with $\alpha(L) = \mu$?

2-component Links

In this section we shall specialize to the case $\mu = 2$, and consider there the Σ-localized pairing. As the coefficient ring is then a P.I.D. the algebra simplifies greatly. Firstly $\hat{t}H_1(X')_\Sigma = tH_1(X')_\Sigma$ and the Blanchfield pairing is perfect, for the Universal Coefficient spectral sequence and Poincare duality then give an isomorphism.

$$\mathrm{Hom}_{\Lambda_\Sigma}(tH_1(X')_\Sigma, K/\Lambda_\Sigma) \approx \mathrm{Ext}^1_{\Lambda_\Sigma}(H_1(X')_\Sigma, \Lambda_\Sigma) \rightarrow tH_1(X')_\Sigma$$

which is just the inverse of the adjoint map of the Blanchfield pairing. (In fact a more careful examination of the spectral sequence over $Q\Lambda_S$ shows that the Blanchfield pairing on $Q \otimes tH_1(X')_S$ is already perfect).

Secondly if M is a finitely generated torsion module over a P.I.D. R then $M = \bigoplus_{\mathfrak{p}} M_{\mathfrak{p}}$ (summation over nonzero primes $\mathfrak{p}$ of R) where $M_{\mathfrak{p}} = \bigoplus_{1 \leq i \leq \nu(\mathfrak{p})} (R/\mathfrak{p}^{e_i})$ is the $\mathfrak{p}$-coprimary submodule, and $M_{\mathfrak{p}} \to M \otimes_R R_{\mathfrak{p}}$ is an isomorphism. If b is an ε-linking pairing on M, then it is easily seen that if $\mathfrak{r}, \mathfrak{s}$ are primes such that $\mathfrak{s} \neq \bar{\mathfrak{r}}$, then $M_{\mathfrak{r}}$ and $M_{\mathfrak{s}}$ are orthogonal, that is $b(M_{\mathfrak{r}}, M_{\mathfrak{s}}) = 0$. Hence M is an orthogonal direct sum

$$M = \left(\bigoplus_{\mathfrak{q}=\bar{\mathfrak{q}}} M_{\mathfrak{q}}\right) \oplus \left(\bigoplus_{\mathfrak{r} \neq \bar{\mathfrak{r}}} (M_{\mathfrak{r}} \oplus M_{\bar{\mathfrak{r}}})\right) .$$

If $\mathfrak{r} \neq \bar{\mathfrak{r}}$ then $M_{\mathfrak{r}}^{\perp} = M_{\mathfrak{r}}$ in $M_{\mathfrak{r}} \oplus M_{\bar{\mathfrak{r}}}$, so the second big summand is neutral. This splitting of linking pairings induces a map

$$W_{\varepsilon}(F,R,-) \xrightarrow{\sim} \bigoplus_{\mathfrak{q}=\bar{\mathfrak{q}}} W_{\varepsilon}(F,R_{\mathfrak{q}},-) .$$

Furthermore, since $R_{\mathfrak{q}}$ is a discrete valuation ring, $W_{\varepsilon}(F,R_{\mathfrak{q}},-) = W_{\varepsilon}(k_{\mathfrak{q}},-)$, the Witt group of nonsingular ε-Hermitian forms over the residue field $k_{\mathfrak{q}} = R_{\mathfrak{q}}/\mathfrak{q} = R/\mathfrak{q}$ of $R_{\mathfrak{q}}$. This is a standard result, which may be proved in the following way. There is a natural monomorphism $W_{\varepsilon}(k_{\mathfrak{q}},-) \to W_{\varepsilon}(F,R_{\mathfrak{q}},-)$, since a vector space over $k_{\mathfrak{q}}$ may be regarded as an R-torsion module and a nonsingular ε-Hermitian form on such a vector space may be regarded as an ε-linking pairing with values in the submodule $\mathfrak{q}^{-1}R/R \approx k_{\mathfrak{q}}$ of K/R. That it is onto follows from the following lemma (also well-known):

Lemma 7 If b is a primitive ε-Hermitian pairing on a finitely generated R-torsion module M, and if N is a finitely generated submodule of M such that $N \subset N^{\perp}$ then (M,b) is Witt-equivalent to $(N^{\perp}/N, b_N)$, where b_N is the primitive ε-Hermitian pairing induced on $N^{\perp}/N$ by b.

Proof The induced pairing b_N on $N^{\perp}/N$, defined by

$b_N(n_1 \bmod N, n_2 \bmod N) = b_N(n_1,n_2)$ for all n_1, n_2 in $N^{\perp}$ is clearly a primitive ε-Hermitian pairing. Let $P = \{\langle p, p \bmod N\rangle \mid p \text{ in } N^{\perp}\}$ be the image of $N^{\perp}$ in $M \oplus (N^{\perp}/N)$ via the diagonal embedding. Then P is selforthogonal with respect to the pairing $b \oplus (-b_N)$, for if $b \oplus (-b_N)(\langle m, n \bmod N\rangle, \langle p, p \bmod N\rangle) = 0$ for all p in N , where m is in M and n is in $N^{\perp}$, then $b(m-n, p) = 0$ for all p in $N^{\perp}$, and so m-n is in N. In particular m is in $N^{\perp}$ and $\langle m, n \bmod N\rangle = \langle m, m \bmod N\rangle$ is in P. Therefore $(M,b) \oplus (N^{\perp}/N, -b_N)$ is neutral. //

Now if (M,b) is an ε-linking pairing over a discrete valuation ring $R_{\mathfrak{q}}$ and M is annihilated by $\mathfrak{q}^m$ where $m > 1$, then b takes values in the submodule $\mathfrak{q}^{-m}R/R$ of K/R. Hence $N = \mathfrak{q}^{m-1}M$ is such a submodule of M (for $b(N,N) \subset \mathfrak{q}^{2m-2}(\mathfrak{q}^{-m}R/R) = 0$) and $N^{\perp}/N$ is annihilated by $\mathfrak{q}^{m-1}$, so on using the lemma repeatedly, (M,b) is Witt-equivalent to some (M'',b') where $\mathfrak{q}M' = 0$. Clearly M' is then a vector space over $k_{\mathfrak{q}}$, and b' takes values in $\mathfrak{q}^{-1}R/R \approx k$, so may be regarded as an ε-Hermitian form on M'. Thus $W_{\varepsilon}(F,R,-) = \bigoplus_{\mathfrak{q}=\bar{\mathfrak{q}}} W_{\varepsilon}(k_{\mathfrak{q}},-)$ where $k_{\mathfrak{q}} = R/\mathfrak{q}$ and the summation is over all primes of R left invariant under the involution.

The ring Λ_{Σ} is a localization of the ring $Q(t_1)[t_2]$ of polynomials in one variable t_2 with coefficients in the field $Q(t_1)$, obtained by inverting all polynomials in t_2 with constant coefficients, and is thus a principal ideal domain. It may also be described as the subring of $K = Q(t_1,t_2)$ generated by $Q(t_1) \cup Q(t_2)$. Thus a prime ideal in Λ_{Σ} is generated by an irreducible element $p(t_1,t_2)$ in $Q[t_1,t_2]$ of positive degree in each variable, and those invariant under the involution correspond to such p for which also $p(t_1^{-1},t_2^{-1}) = \alpha(t_1)\beta(t_2)p(t_1,t_2)$ for some units $\alpha(t_1)$ in $Q(t_1)$, $\beta(t_2)$ in $Q(t_2)$. In fact it follows easily from the irreducibility of p and that it involves each variable, that $\alpha(t_1)\beta(t_2) = rt_1^m t_2^n$ for some r in Q and m,n in $\mathbb{Z}$; furthermore since it must also be true that $p(t_1, t_2) = \alpha(t_1^{-1})\beta(t_2^{-1})p(t_1^{-1}, t_2^{-1})$, r^2 must equal 1. Thus $p(t_1^{-1},t_2^{-1}) = \pm t_1^m t_2^n p(t_1,t_2)$. The residue field of such a prime ideal is an algebraic extension of $Q(t_1)$, $k = Q(t_1)[y]$ say (where y satisfies $p(t_1,y) = 0$), and admits a non-trivial involution generated by $t_1 \to t_1^{-1}$, $y \to y^{-1}$. The elment $\alpha = t_1 - t_1^{-1}$ satisfies $\bar{\alpha} = -\alpha$ and consequently $k = k_0[\alpha]$ where $k_0 = \{f \text{ in } k \mid f = \bar{f}\}$ is the fixed field of the involution, by Galois theory. Hence also if β is an ε-Hermitian form on the k-vector space V, then $\alpha\beta$ is an $(-\varepsilon)$-Hermitian form on V, and so in discussing Witt groups of such fields, it suffices to consider $\varepsilon = +1$, which anyway is the case relevant for the application to classical links. Any (+1)-Hermitian form (V, B) may be diagonalized, that is to say V splits as an orthogonal direct sum of 1-dimensional subspaces, and so there is an epimorphism $\mathbb{Z}[k_0^{\cdot}/N_{k/k_0}k^{\cdot}] \twoheadrightarrow W_{+1}(k,-)$ [97].

Not every prime ideal of Λ_Σ invariant under the involution is thus associated to the Alexander module of a boundary link; there is an 'integrality' condition that must be satisfied. In general, for L a μ-component boundary link, $E_\mu(L)$ is principal, generated over Λ_μ by an element Δ_μ such that $\varepsilon(\Delta_\mu) \equiv \Delta_\mu(1, \ldots, 1)$ equals 1 and $\bar{\Delta}_\mu = \Delta_\mu$. It follows that $E_0(tH_1(X')_\Sigma)$ is generated over Λ_Σ by Δ_μ. Now if $M = \oplus R/\mathfrak{p}_i^{e_i}$ is a torsion module over a P.I.D. R then $E_0(M) = \Pi\mathfrak{p}_i^{e_i}$ and thus the primes occurring in the direct sum decomposition are just the divisors of $E_0(M)$. Assuming once again that $\mu = 2$, if $\Lambda_\Sigma/\mathfrak{p}^e$ is a direct summand of $tH_1(X')_\Sigma$ with $\mathfrak{p} = \bar{\mathfrak{p}}$, then $\mathfrak{p}$ is generated by some p in Λ_2 such that $\bar{p} = \pm t^m t^n p$ and $p(1, 1) = 1$, for by the Gauss Content lemma [ ] any factorization of Δ_2 into irreducibles in Λ_Σ comes from a factorization into irreducibles in Λ_2.

On the other hand, if q in Λ_2 is such that $q = \bar{q}$ and $q(1,1) = 1$ then according to Gutiérrez there is a 2-component boundary link L with $\Delta_2(L) = q$ [62]. (Alternatively, Bailey's theorem implies that the matrix $\begin{pmatrix} 0 & 0 \\ 0 & q \end{pmatrix}$ is a presentation matrix for $H_1(X')$ for some 2-component link [7]). In particular, if q is irreducible, then $tH_1(X')_\Sigma = \Lambda_\Sigma/(q)$ is of length 1, and so $B_\Sigma(L)$ maps to a nonzero element of $W_{+1}(k_{\mathfrak{q}},-)$. Thus the image of the set of all 2 component boundary links in $W_{+1}(K_2,\Lambda_\Sigma,-)$ is contained in

$$\bigoplus_{\mathfrak{p}} \left\{ W_+(k_{\mathfrak{p}},-) \mid \mathfrak{p} = (p) \text{ for some } p \text{ in } \Lambda_2 \text{ such that } p(1,1) = 1, \; p \text{ is irreducible and } \bar{p} = \pm X_1^m X_2^n p \right\}$$

and maps non trivially to infinitely many factors of this direct sum, and so is not contained in any finitely generated subgroup.

<u>A late addition</u>: Levine has announced work on the (unlocalized) Blanchfield pairing on $A = H_1(X;\Lambda)$ for $\mu = 2$ [120a]. If $\alpha(L) = 1$ the kernel and cokernel of the adjoint map from A to Hom $(\bar{A},Q(t)/\Lambda)$ depend only on the linking number; if $\alpha(L) = 2$ they are determined by longitude-annihilating polynomials $b_1(t_2)$ and $b_2(t_1)$ (compare our Theorem VI.4) and an ideal I in Λ_2 such that $b_1(1) = b_2(1) = 1$, $b_1(t_2) + b_2(t_1) - 1$ is in I and $\varepsilon(I) = \mathbb{Z}$. Moreover $A/tA \approx I$ and, using Bailey's thesis, any such triple b_1, b_2, I may be realized.

Signatures

An argument similar to the one of the preceding section shows that (for $\mu = 1$) there is an isomorphism $W_\varepsilon(\mathbb{R}(t), \mathbb{R}\Lambda, -) \approx \bigoplus W_\varepsilon(\mathbb{R}(t)/p(t), -)$, where the sum is taken over all irreducible real polynomials such that $(p(t)) = (p(t^{-1}))$. Apart from $t + 1$ and $t - 1$, which play no role in knot theory, any such polymomial must be a quadratic of the form $p_\theta(t) = t^2 - \cos\theta . t + 1$, for some $0 \leqslant \theta \leqslant \pi$. The induced maps of $W_\varepsilon(Q(t),\Lambda, -)$ onto $W_\varepsilon(\mathbb{R}(t)/p_\theta(t), -) \approx \mathbb{Z}$ are essentially the Milnor signatures σ_θ [95, 132]. Murasugi and Tristram have defined signatures for any classical link, and have shown them to be concordance invariants [141, 193]. Certain of these signatures have been reinterpreted by Kauffman and Taylor, and by Viro, who applied the G-index theorem to branched cyclic covers of D^4, branched over a properly embedded spanning surface for the link [88, 201]. A simple algorithm for the Murasugi signature has been given by Gordon and Litherland [60]. The relation between the Milnor signatures and the Tristram signatures of an Hermitean pairing have been elucidated by Matumoto [126]. All of these signatures appear to be related to (finite, branched) cyclic covers of the link. Cooper has defined "multisignatures" for 1-links analogous to the Tristram signatures, but apparently more closely tied to the structure of the maximal abelian cover [34]. He observes that his invariants vanish for slice links, but can be nonzero even if all the Tristram signatures are 0. Can all of these signatures be interpreted as homomorphisms from $W_+(K_\mu,\Lambda_\mu, -)$ to $\mathbb{Z}$?

Appendix: A Surgical View of the Blanchfield Pairing

In this section we shall describe the Blanchfield pairing of a 1-knot by means of the surgery technique of Hempel [68], Levine [115], and Rolfsen [156]. In the latter papers this technique was used to characterize the Alexander polynomials of a knot. Bailey used it to prove the theorem quoted in Chapter VII [7]. Our construction is a reworking of his argument, paying closer attention to the Blanchfield pairing; indeed our ultimate goal is that we may better understand his theorem, and extend it to a characterization of the Blanchfield pairing of 2-component links.

It is well known that a 1-knot K may be unknotted by "replacing certain of the overcrossings by undercrossings"; this idea is made precise in the following lemma of Hempel [68]. (We have refined the statement, as in Rolfsen [157; page 159] and Bailey [7; page 26]).

Lemma (Hempel) There is an embedding $L : mS^1 \times D^2 \longrightarrow X$ and an homeomorphism h of $U = S^3 - L(mS^1 \times \text{int}D^2)$ onto itself such that

(i) the induced map from $mS^1 \times \{0\}$ to S^3 is a trivial link, whose i^{th} longitude and meridian are represented by the images of $S^1 \times \{1\} \times \{i\}$ and $\{1\} \times S^1 \times \{i\}$ respectively, for $1 \leq i \leq m$;

(ii) h maps each boundary component of U to itself and (on the ith meridian) $h(L(1,s,i)) = L(s,s,i)$ for all s in S^1 and $1 \leq i \leq m$;

(iii) each component of this link has linking number 0 with K and with h(K);

(iv) hoK is unknotted in S^3.

Let $\phi : S^1 \times S^1 \longrightarrow \partial X$ be an homeomorphism carrying $\{1\} \times S^1$ and $S^1 \times \{1\}$ to a longitude and a meridian of K respectively, and let $Y = X \cup_\phi S^1 \times D^2$ be the result of 0-framed surgery on K. Let $T = mS^1 \times D^2$,

$V = \overline{X - L(T)}$ and $W = V \cup_\phi S^1 \times D^2 = \overline{Y - T}$. Let $g = h^{-1} \circ L \mid \partial T$. Then the pair $(U \cup_g T, K(S^1))$ is homeomorphic to the pair $(S^3, h\circ K(S^1)) = (U \cup T, h\circ K(S^1))$ via the map given by h on U and by the identity on T. Therefore $W \cup_g T = (V \cup_g T) \cup_\phi S^1 \times D^2$ is homeomorphic to $Z = S^1 \times S^2$, since it is the result of 0-framed surgery on the unknot K in $U \cup_g T \approx S^3$. Let e and f be the induced embeddings of T and W (respectively) in Z. The inclusion of the meridian of K into each of X, Y, U, W and $W \cup_g T = Z$ induces isomorphisms on the first integral homology groups, and hence identifies each of these groups with $\mathbb{Z}$. We shall denote the maximal abelian covering spaces of these spaces (and the induced coverings of subspaces and lifts of maps) by affixing a prime ' . The homology groups of such covering spaces are then finitely generated Λ_1-modules. (All unlabelled maps are induced by inclusions).

Since $Z' = \mathbb{R} \times S^2 = S^3 - S^0$, $H_1(Z;\Lambda) = 0$ and $H_2(Z;\Lambda) = \mathbb{Z}$, while since $T' = T \times \mathbb{Z}$, $H_1(T;\Lambda) = \Lambda^m = H_2(T,\partial T;\Lambda)$ and $H_1(T,\partial T;\Lambda) = H_2(T;\Lambda) = 0$. Since X' may be constructed by splitting X along a Seifert surface, the longitude of K lifts to a loop in X' which is null homologous there (cf. Chapter I) and so the natural map from $H_1(X;\Lambda)$ to $H_1(Y;\Lambda) = H_1(X' \cup \mathbb{R} \times D^2;\mathbb{Z})$ is an isomorphism, while $H_2(Y;\Lambda) = H_2(X;\Lambda) \oplus \mathbb{Z} = \mathbb{Z}$. By excision, $H_2(Y,W;\Lambda) = H_2(T,\partial T;\Lambda) = \Lambda^m$ and $H_1(Y,W;\Lambda) = H_1(T,\partial T;\Lambda) = 0$, so the long exact sequence of the pair (Y',W') gives a short exact sequence

$$0 \longrightarrow H_2(Y,W;\Lambda) \xrightarrow{D} H_1(W;\Lambda) \xrightarrow{\psi} H_1(Y;\Lambda) \longrightarrow 0$$

as the natural map from $H_2(Y;\Lambda) = \mathbb{Z}$ to $H_1(Y,W;\Lambda) = \Lambda^m$ must be null. Similarly in the Mayer-Vietoris sequence of equivariant homology for (Z',W',T') the boundary map from $H_2(Z;\Lambda)$ to $H_1(\partial T;\Lambda)$ is null, so there is an isomorphism from Ker $(: H_1(\partial T;\Lambda) \longrightarrow H_1(T;\Lambda))$ to $H_1(W;\Lambda)$, which is therefore free of rank m.

As generators for $H_1(W;\Lambda)$ we may take the Alexander duals in $Z' \subset S^3$ of the images under e' of fixed lifts to T' of the cores of the components of T. Thus if $c_1,\ldots,c_m$ are such lifts, whose images generate $H_1(T;\Lambda)$, the dual basis of $H_1(W;\Lambda)$ is determined by 1-cycles $a_1,\ldots,a_m$ such that

$$\ell(t^m f'_* a_i,\ t^n e'_* c_j) = 1 \text{ if } m = n \text{ and } i = j,$$
$$= 0 \text{ otherwise.}$$

(Here $\ell(\alpha,\beta)$ denotes the linking number of two disjoint 1-cycles α and β in S^3). The module $H_2(Y,W;\Lambda)$ is generated by the images under L' of discs in T' transverse to the cores $c_1,\ldots, c_m$. Let d_i be the ith such disc in Y', and let $D(d_i) = \sum_{n\in\mathbb{Z}} \sum_{1\leq j\leq m} r_{ijn}t^n a_j$ be its image in $H_1(W;\Lambda)$. Since $D(d_i)$ is represented by ∂d_i, it follows that $r_{ijn} = \text{link }(f'_*\, d_i,\ t^n c_j)$, and since $f'_*\, d_i = g(1\times S^1\times\{i\})$ is homologous to $e'_* c_i$ in $e'(T')$, $r_{ijn} = r_{ji-n}$. In other words if $\mathbb{D}$ is the matrix of D with respect to the bases d_i and a_j , then $\mathbb{D} = \bar{\mathbb{D}}^{tr}$. Furthermore since the sequence above is exact (or since $H_1(Y;\Lambda) = H_1(X;\Lambda) = tA(K)$ is a torsion module), $\delta = \det \mathbb{D}$ is nonzero.

We claim that the Blanchfield pairing on $H_1(Y;\Lambda)$ is given by

$$[\psi(\Sigma u_i a_i),\ \psi(\Sigma v_j a_j)] = \bar{v}^{tr}\mathbb{D}^{-1}u \text{ modulo } \Lambda.$$

(Here (u_i) and $v_j)$ are regarded as column vectors in Λ^μ and so the matrix product on the right lies in $Q(t)$).

Let z be a 1-cycle on W', and S a 2-chain on Y' such that $\partial S \subset W'$. Then $I_{Y'}(z,\ S) = I_{W'}(z,\ S\cap W') = I_{S^3}(f'_*(z),\ f'_*(S\cap W'))$ since f' embeds W' into $Z' \subset S^3$. This in turn equals $\text{link }(f'_*z,\ \partial f'_*(S\cap W')) = \text{link}(f'_*z,\ f'_*\partial S) + \text{link }(f'_*z,\ f'_*\ (S\cap\partial W'))$. Since $\delta H_1(Y';\Lambda) = 0$, there is a 2-chain S_j on Y' such that $\partial S_j = \delta a_j$.

Since δa_j is homologous to $D\left(\sum_{1\leq k\leq m}\left(\delta\mathbb{D}^{-1}\right)_{kj}\, d_k\right)$, $S \cap \delta W'$ is homologous to $\sum_{1\leq k\leq m} (\delta\mathbb{D}^{-1})_{kj}\, \partial\, d_K$ in W'. Therefore

$$[\psi(a_i),\ \psi(a_j)] = \frac{1}{\bar{\delta}} \sum_{n\in\mathbb{Z}} I_{Y'}(a_i,\ t^n S_j)t^n \text{ modulo } \Lambda$$

$$= \frac{1}{\bar{\delta}} \sum_{n\in\mathbb{Z}} \text{link}(f'_* a_i,\ t^n \delta f'_* a_j)t^n +$$

$$\frac{1}{\bar{\delta}} \sum_{n\in\mathbb{Z}} \text{link}(f'_* a_i,\ t^n \sum_{1\leq k\leq m} (\delta\mathbb{D}^{-1})_{kj}\, f'_*\, \partial\, d_k)t^n \text{ modulo } \Lambda$$

$$= \sum_{n\in\mathbb{Z}} \sum_{1\leq k\leq m} \frac{\overline{(\delta\mathbb{D}^{-1})}}{\bar{\delta}}_{kj} \text{link}(f'_* a_i,\ t^n f'_*\, \partial\, d_k)t^n \text{ modulo } \Lambda$$

$$= \sum_{n\in\mathbb{Z}} \sum_{1\leq k\leq m} (\mathbb{D}^{-1})_{kj}\ \text{link}(f'_* a_i,\ t^n e'_* c_k)t^n \text{ modulo } \Lambda$$

$$= (\mathbb{D}^{-1})_{ij} \text{ modulo } \Lambda$$

which establishes our claim. (In simplifying the R.H.S. we have used the facts that $\sum_{n\in\mathbb{Z}} \text{link}\,(\alpha,\ t^n\beta)t^n$ is Λ-sesquilinear in α and β, and that $\mathbb{D}$ is Hermitean).

CHAPTER X NONORIENTABLE SPANNING SURFACES

In the course of studying the quadratic form on the total linking number infinte cyclic cover of a link complement, Murasugi defined the nullity of a link L, $\eta(L)$, to be $1 + \text{nullity}(V + V^{tr})$ where V is the Seifert matrix for any connected spanning surface for L [138]. He showed that $\eta(L)$ lay between 1 and μ, and was equal to the nullity of the Jacobian matrix evaluated at $(-1,\ldots,-1)$, and Kauffman and Taylor, in a rederivation of his results, showed that $\eta(L)$ is invariant under I-equivalence [88]. The Alexander nullity shares some of these properties, and is defined in a similar fashion. In this chapter it shall be shown that, nevertheless, the two invariants are distinct, and can take any values such that $1 \leq \alpha(L) \leq \eta(L) \leq \mu$. Our examples are constructed from links spannable by disjoint nonorientable surfaces, and we prove an analogue of Smythe's theorem characterizing boundary links. We show also that a μ-component link which is so spannable must have Murasugi nullity μ, and that this condition is sufficient if $\mu = 2$, and we give a geometric interpretation for the invariance of the Murasugi nullity under concordance, in the 2-component case.

Figure 1

<u>Definition</u> A μ-component link $L : \mu S^1 \to S^3$ is a <u>$\mathbb{Z}/2\mathbb{Z}$-boundary link</u> if there is an embedding $P : \coprod_{1 \leq i \leq \mu} U_i \to S^3$ of μ disjoint surfaces U_i such that $L_i = P|\partial U_i$.

The surfaces are not required to be orientable, in contrast to the definition of boundary links. For instance, the link depicted in figure 1 is clearly spanned by two Möbius bands and so is a $\mathbb{Z}/2\mathbb{Z}$-boundary link, but it is very far from being a boundary link, for it has linking number 4. As this link has $\alpha = 1$ and $\eta = 2$, it already suffices to show that the Alexander and Murasugi nullities are not always equal. (This is the link 9^2_{61} in the tables of [157]). The link of figure V.1 is a more subtle example, for it has Alexander polynomial zero, and one of the disjoint spanning surfaces may be assumed orientable, yet is is not an homology boundary link. There is a characterization of $\mathbb{Z}/2\mathbb{Z}$-boundary links in terms of the link group, analogous to that given for boundary links by Smythe.

<u>Theorem 1</u> <u>A μ-component link L is a $\mathbb{Z}/2\mathbb{Z}$-boundary link if and only if there is a map $f : G \to \overset{\mu}{*}(\mathbb{Z}/2\mathbb{Z})$ which carries some i^{th} meridian to the generator of the ith factor of the free product, for each $1 \leq i \leq \mu$.</u>

<u>Proof</u> Assume first that L is a $\mathbb{Z}/2\mathbb{Z}$-boundary link with spanning surfaces U_i. Each such surface has an open regular neighbourhood homeomorphic to the total space of its normal bundle ν_i in X. Crushing the complement of a disjoint family of such neighbourhoods to a point collapses X onto the wedge of Thom spaces [85; page 204], $X \to \bigvee_{i=1}^{\mu} T(\nu_i)$. The normal bundles ν_i are induced from the canonical line bundle η_N over $\mathbb{R}P^N$ (for N large) by classifying maps $n_i : U_i \to \mathbb{R}P^N$, and these maps induce a map

$T(\eta) : \overset{\mu}{\vee} T(\nu_i) \to \overset{\mu}{\vee} T(\eta_N)$. Now $T(\eta_N)$ is homeomorphic to $\mathbb{R}P^{N+1}$ by a homeomorphism carrying the zero section to the hyperplane at infinity [85; page 205], and the inclusion of $\mathbb{R}P^{N+1}$ into $\mathbb{R}P^{\infty} = K(\mathbb{Z}/2\mathbb{Z},1)$ is an (N+1)-connected map, and so the inclusion of $\overset{\mu}{\vee} \mathbb{R}P^{N+1}$ into $\overset{\mu}{\vee} \mathbb{R}P^{\infty} = K(\overset{\mu}{*}(\mathbb{Z}/2\mathbb{Z}),1)$ is (N+1)-connected. The composition of these maps gives a map $X \to K(\overset{\mu}{*}(\mathbb{Z}/2\mathbb{Z}),1)$ which determines a map $f : G \to \overset{\mu}{*}(\mathbb{Z}/2\mathbb{Z})$. This map carries an ith meridian represented by a loop from the basepoint which meets U_i transversally in one point and is disjoint from the other surfaces U_j to the generator of the ith factor. For the map of spaces $X \to \overset{\mu}{\vee} \mathbb{R}P^{N+1}$ carries such a meridianal curve onto the Thom space of the restriction of η_N to a point, in other words to a curve which intersects the zero section hyperplane $\mathbb{R}P^N$ in one point, and hence which is essential in $T(\eta_N) = \mathbb{R}P^{N+1}$, and so in $\mathbb{R}P^{\infty}$ (since the inclusion $\mathbb{R}P^{N+1} \to \mathbb{R}P^{\infty}$ is 2-connected).

Conversely, given a map $f : G \to \overset{\mu}{*} \mathbb{Z}/2\mathbb{Z}$, it may be realized by a map $F : X \to \overset{\mu}{\vee} \mathbb{R}P^{\infty}$, since the latter space is an Eilenberg-MacLane space $K(\overset{\mu}{*}(\mathbb{Z}/2\mathbb{Z}),1)$. By general position it may be assumed that f maps X to $\overset{\mu}{\vee} \mathbb{R}P^N$ with N large. If f maps an ith meridian to the generator of the ith $\mathbb{Z}/2\mathbb{Z}$ factor, then it must map the corresponding longitude to the identity (since it commutes with this meridian, but lies in the commutator subgroup modulo the meridians of the other components). It may then be assumed that $F|\partial X_i$ maps a longitudinal curve into the hyperplane $\mathbb{R}P^{N-1}$ of the corresponding $\mathbb{R}P^N$, and is transverse to that hyperplane. On moving F transverse to the union of these hyperplanes $\coprod^{\mu} \mathbb{R}P^{N-1}$ (which may be assumed disjoint from the sole singularity, the wedge point, and hence from each other) the inverse image $F^{-1}(\coprod^{\mu} \mathbb{R}P^{N+1})$ is a family of disjoint surfaces spanning the link. //

In the original theorem of Smythe characterizing boundary links the spanning surfaces had trivial normal bundles, and the universal trivial line bundle $\mathbb{R}$ (with base space a point) has Thom space $T(\mathbb{R}) = S^1$, which there played the role which $T(\eta_N) = \mathbb{R}P^{N+1}$ plays in the above theorem. A nonorientable surface is a $\mathbb{Z}/2\mathbb{Z}$-manifold in the sense of [134]. A similar application of transversality to high dimensional lens spaces shows that L has μ disjoint spanning complexes, the ith being a $\mathbb{Z}/p_i\mathbb{Z}$-manifold with no singularities on the boundary, if and only if there is a map $G \to \underset{1 \leq i \leq \mu}{*} (\mathbb{Z}/p_i\mathbb{Z})$ carrying some i^{th} meridian to a generator of the ith factor. (For there is a 2N-dimensional $\mathbb{Z}/p\mathbb{Z}$-manifold in $L_N(p) = S^{2N+1}/(\mathbb{Z}/p\mathbb{Z})$ whose homology class generates $H_{2N}(L_N(p); \mathbb{Z}/p\mathbb{Z})$, the Poincaré dual of $H^1(L_N(p); \mathbb{Z}/p\mathbb{Z})$ [31; page 89]. The relevance of $\mathbb{Z}/p\mathbb{Z}$-manifolds to knot theory was observed by Cooper.)

Finite dimensional approximations $\mathbb{R}P^N$ to $\mathbb{R}P^\infty$ have been used to facilitate the distinction between the base space ($\mathbb{R}P^N$) and the Thom space ($\mathbb{R}P^{N+1}$) of the universal line bundle. Since X has the homotopy type of a 2-dimensional complex, any $N \geq 2$ would suffice. (Note that similarly the inclusion of $L_N(p)$ in $L_\infty(p) = K(\mathbb{Z}/p\mathbb{Z}, 1)$ is highly connected.)

Smythe's characterization of homology boundary links suggests two possible definitions for a $\mathbb{Z}/2\mathbb{Z}$-homology boundary link.

Definition (a) A μ-component link L is a $\mathbb{Z}/2\mathbb{Z}$-homology boundary link if there are μ disjoint surfaces (not necessarily orientable) U_i in $X(L)$ with $\partial U_i \subset \partial X(L)$ and such that ∂U_i is $\mathbb{Z}/2\mathbb{Z}$-homologous to the ith longitude in $\partial X(L)$.

(b) A μ-component link L is a weak $\mathbb{Z}/2\mathbb{Z}$-homology boundary link if there is an epimorphism $G \to \overset{\mu}{*}(\mathbb{Z}/2\mathbb{Z})$.

The above theorem then has the following analogue (with a similar proof, which we shall not give).

Theorem 2 A link L is a $\mathbb{Z}/2\mathbb{Z}$-homology boundary link if and only if there is an epimorphism $G \to \overset{\mu}{*}(\mathbb{Z}/2\mathbb{Z})$ such that the composition $G \to \overset{\mu}{*}(\mathbb{Z}/2\mathbb{Z}) \to \overset{\mu}{\oplus}(\mathbb{Z}/2\mathbb{Z})$ carries the ith meridian to the generator of the ith summand. //

The last clause is superfluous in the case of homology boundary links, for any automorphism of $\mathbb{Z}^{\mu}$ can be lifted to an automorphism of $F(\mu) = \overset{\mu}{*}\mathbb{Z}$ [123; page 168] and so any epimorphism of G onto $F(\mu)$ can be changed so as to carry meridians to standard generators of the abelianization.

Theorem 3 The Murasugi nullity of a μ-component $\mathbb{Z}/2\mathbb{Z}$-homology boundary link is μ.

Proof Let L be a μ-component link with group G. The total $\mathbb{Z}/2\mathbb{Z}$ linking number homomorphism from G to $\mathbb{Z}/2\mathbb{Z} = \{\pm 1\}$, which sends each meridian to -1, determines an epimorphism from $\Lambda = \mathbb{Z}[G/G']$ to $\tilde{\mathbb{Z}} = \Lambda/(t_1+1,\ldots,t_\mu+1)$. There is a canonical isomorphism $\mathbb{Z}[\overset{\mu}{\oplus}(\mathbb{Z}/2\mathbb{Z})] = \Lambda/(t_1^2-1,\ldots,t_\mu^2-1)$, so $A(\overset{\mu}{*}(\mathbb{Z}/2\mathbb{Z}))$ has a canonical Λ-module structure, and the map sending each standard generator of $\overset{\mu}{\oplus}(\mathbb{Z}/2\mathbb{Z})$ to -1 in $\{\pm 1\}$ induces the projection of $\mathbb{Z}[\overset{\mu}{\oplus}(\mathbb{Z}/2\mathbb{Z})]$ onto $\tilde{\mathbb{Z}}$. Hence if there is an epimorphism $f : G \to \overset{\mu}{*}(\mathbb{Z}/2\mathbb{Z})$ such that the composition with the map sending the generator of each factor to -1 in $\{\pm 1\}$ is the total $\mathbb{Z}/2\mathbb{Z}$-linking number homomorphism, then the epimorphism $A(f) : A(G) \to A(\overset{\mu}{*}(\mathbb{Z}/2\mathbb{Z}))$ gives rise to an epimorphism $\tilde{\mathbb{Z}} \otimes_\Lambda A(G) \to \tilde{\mathbb{Z}} \otimes_\Lambda A(\overset{\mu}{*}(\mathbb{Z}/2\mathbb{Z})) = \tilde{\mathbb{Z}}^{\mu}$. Since this is the case for a $\mathbb{Z}/2\mathbb{Z}$-homology boundary link, and since the Jacobian matrix evaluated at $(-1,\ldots,-1)$ is a presentation matrix for $\tilde{\mathbb{Z}} \otimes_\Lambda A(G)$ over $\tilde{\mathbb{Z}}$, the theorem follows. //

Corollary Given integers $1 \leq \alpha \leq \eta \leq \mu$, there is a μ-component link with Alexander nullity α and Murasugi nullity η.

Proof Let $\mu' = \mu - \alpha + 1$ and $\eta' = \eta - \alpha + 1$. Let L_0 be a μ'-component link all of whose pairwise linking numbers are odd, and let L' be the link obtained by replacing the ith component of L_0 by its (2,1)-cable (that is, by the boundary of a Möbius band whose core is that component) for $1 \leq i \leq \eta'$. Then $\alpha(L') = 1$ since all the linking numbers of L' are nonzero. (This follows from the second Torres conditions, used inductively to conclude that $\Delta_1(L)(t_1, 1, \ldots, 1) \neq 0$.) The link obtained by deleting the first $(\eta' - 1)$ components of L' has Murasugi nullity 1, so $\eta(L') \leq 1 + (\eta' - 1) = \eta'$. On the other hand the group of L' maps onto $\overset{\eta'}{*}(\mathbb{Z}/2\mathbb{Z})$ and so an argument as in the theorem shows that $\eta(L') \geq \eta'$. Therefore $\eta(L') = \eta'$. Let L'' be a trivial $(\alpha - 1)$-component link and let $L = L' \coprod L''$ be the disjoint union (so that L' and L'' are separated by a 2-sphere in S^3). Then $\alpha(L) = \alpha$, $\eta(L) = \eta$ and $\mu(L) = \mu$, since these invariants are clearly additive for such disjoint unions. (If $\alpha \geq 2$ we may construct an example more simply by taking the disjoint union of an η'-component $\mathbb{Z}/2\mathbb{Z}$-boundary link L_1 with all linking numbers nonzero, a $(\mu - \eta + 1)$-component link L_1 with Hosokawa polynomial $\nabla(L_2) = 1$ (so that $\eta(L_2) = 1$) and an $(\alpha - 2)$-component trivial link.) //

We may consider also the reduced nullity and the sequence $1 \leq \alpha(L) \leq \kappa(L) \leq \eta(L) \leq \mu$. Are the members of this sequence independent?

The proof of the theorem actually shows that if L is a weak $\mathbb{Z}/2\mathbb{Z}$-homology boundary link and if the epimorphism $f : G \to \overset{\mu}{*}(\mathbb{Z}/2\mathbb{Z})$ is such that composition with the map sending the generator of each factor of $\overset{\mu}{*}(\mathbb{Z}/2\mathbb{Z})$ to (-1) in $\{\pm 1\}$ is the total $\mathbb{Z}/2\mathbb{Z}$-linking number homomorphism of L, then $\eta(L) = \mu$. It is easily seen that if $\mu = 2$ such a weak $\mathbb{Z}/2\mathbb{Z}$-homology boundary link is a $\mathbb{Z}/2\mathbb{Z}$-homology boundary link. In their discussion of the Murasugi nullity, Kauffman and Taylor used branched coverings associated with the total $\mathbb{Z}/2\mathbb{Z}$-linking number homomorphism.

The boundary of an annulus embedded in S^3 with unknotted core and two full twists is a 2-component link with linking number 2 and group presented by $\{a,x \mid (ax)^2 = (xa)^2\}$ which maps onto $\mathbb{Z} * (\mathbb{Z}/2\mathbb{Z}) = \{u,v \mid v^2 = 1\}$ via the map sending a to u and x to $u^{-1}v$, and hence is a weak $\mathbb{Z}/2\mathbb{Z}$-homology boundary link (as follows from Theorem 4 below). The notions of $\mathbb{Z}/2\mathbb{Z}$-homology boundary link and $\mathbb{Z}/2\mathbb{Z}$-boundary link are probably distinct also, if $\mu \geq 3$, although we know of no examples to show this.

However there is the following pleasant fact. A $\mathbb{Z}/2\mathbb{Z}$-homology boundary link with 2 components is a $\mathbb{Z}/2\mathbb{Z}$-boundary link. This follows from the algebraic characterizations of such links, and is a consequence of Theorems 3 and 4, but can be seen more directly by considering how the Seifert surfaces meet the components of ∂X. In Figure 2 is shown a cross-section through a neighbourhood of the first component of sucha link, and visibly all but one of the (odd number of) boundary components of the Seifert surfaces parallel to this component may be successively paired off and desingularized.

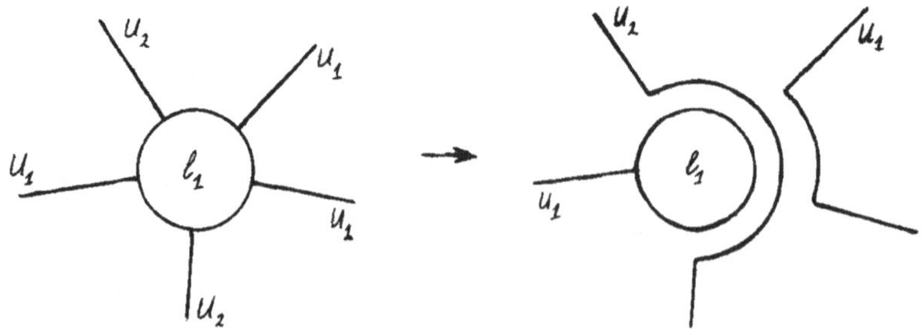

Figure 2

Moreover, whether a 2-component link is a $\mathbb{Z}/2\mathbb{Z}$-boundary link is determined by its Murasugi nullity.

__Theorem 4__ __Let L be a 2-component link. Then the following are equivalent:__

(1) $\eta(L) = 2$,

(2) $\Delta_1(L)(-1,-1) = 0$,

(3) __L is a $\mathbb{Z}/2\mathbb{Z}$-boundary link.__

Moreover if these hold then the linking number of L is divisible by 4.

__Proof__ Let B denote the Λ_2-module G'/G'', and let $\tilde{\mathbb{Z}}$ denote the Λ_2-algebra $\Lambda_2/(t_1+1, t_2+1)$. (As a ring $\tilde{\mathbb{Z}}$ is isomorphic to $\mathbb{Z} = \Lambda_2/(t_1-1,\ t_2-1)$.) Now $E_1(L) = (\Delta_1(L))(t_1-1, t_2-1)$ by Theorem IV.2, while by lemma III.6 $E_1(L) = E_0(B).E_1(I) = E_0(B)(t_1-1, t_2-1)$. Since the Jacobian matrix of G evaluated at $(-1,-1)$ is a presentation matrix for the $\tilde{\mathbb{Z}}$-module $\tilde{\mathbb{Z}} \otimes_\Lambda A(L)$, $\eta(L) = 2$ if and only if $E_1(\tilde{\mathbb{Z}} \otimes A(L)) = 0$ if and only if $\Delta_1(L)(-1,-1) = 0$. Assuming this is the case, then $E_0(\tilde{\mathbb{Z}} \otimes B) = 0$ and so $\tilde{\mathbb{Z}} \otimes B$ maps onto $\tilde{\mathbb{Z}}$. (The kernel is the $\tilde{\mathbb{Z}}$-torsion submodule of $\tilde{\mathbb{Z}} \otimes B$, for $E_1(B) \supseteq E_2(L)$ by Traldi's result of Chapter IV, and $E_2(\mathbb{Z} \otimes A(L))$ is the unit ideal (1) of $\mathbb{Z}$, so $E_2(\tilde{\mathbb{Z}} \otimes A(L))$ must be nonzero, as these ideals have the same image $E_2((\mathbb{Z}/2\mathbb{Z}) \otimes_\Lambda A(L))$ in the common quotient Λ_2-algebra $\mathbb{Z}/2\mathbb{Z}$. Hence $E_1(\tilde{\mathbb{Z}} \otimes B) \neq 0$.) Hence there is a Λ_2-epimorphism $p : B \to \tilde{\mathbb{Z}}$ and $\ker p$, being a Λ_2-submodule of $B = G'/G''$, is a normal subgroup of G/G''. Let $\Gamma = (G/G'')/\ker p$. Then there is a commutative diagram

$$\begin{array}{ccccccccc}
1 & \longrightarrow & G' & \longrightarrow & G & \xrightarrow{a} & \mathbb{Z}^2 & \longrightarrow & 1 \\
 & & \big\downarrow & & \big\downarrow q & & \big\downarrow 1 & & \\
1 & \longrightarrow & \tilde{\mathbb{Z}} & \longrightarrow & \Gamma & \longrightarrow & \mathbb{Z}^2 & \longrightarrow & 1
\end{array}$$

where a carries an ith meridian x_i to the ith standard generator of $\mathbb{Z}^2$

(for $i = 1,2$). Therefore Γ is generated by the images $q(x_1)$, $q(x_2)$ and by a generator of $\widetilde{\mathbb{Z}}$, and the Λ_2-module structure on $\widetilde{\mathbb{Z}} = \Gamma'$ is given by conjugation with the images of the meridians. Therefore Γ has a presentation

$$\{ y_1, y_2, z \mid y_1 z y_1^{-1} = y_2 z y_2^{-1} = z^{-1}, [y_1, y_2] = z^k \}$$

for some k, which clearly must be odd since $\Gamma/\Gamma' = \mathbb{Z}^2$. By choosing the meridians x_i carefully, and on replacing z by its inverse if necessary, we may assume $k = 1$. For suppose that $[y_1, y_2] = z^{4j\pm 1}$, and let w be an element of G such that $q(w) = z$. Let $x_1' = x_1$ and $x_2' = w^j x_2 w^{-j}$. Then x_1' and x_2' are meridians such that $[q(x_1'), q(x_2')]$ generates Γ'. Thus Γ has a presentation

$$\{ y_1, y_2, z \mid z = [y_1, y_2], y_1 z y_1^{-1} = y_2 z y_2^{-1} = z^{-1} \}$$

where the generator y_i corresponds to the element $q(x_i)$, the image of an ith meridian. This presentation is Tietze-equivalent to

$$\{ y_1, y_2 \mid y_1^2 y_2 = y_2 y_1^2 , y_1 y_2^2 = y_2^2 y_1 \}$$

and so the centre of Γ is generated by y_1^2 and y_2^2. Let s be the projection of Γ onto $\Gamma/\text{centre } \Gamma \approx (\mathbb{Z}/2\mathbb{Z}) * (\mathbb{Z}/2\mathbb{Z})$; then $s \circ q : G \to (\mathbb{Z}/2\mathbb{Z}) * (\mathbb{Z}/2\mathbb{Z})$ carries an ith meridian to the generator of the ith factor, and so L is a $\mathbb{Z}/2\mathbb{Z}$-boundary link. The implication (3) $\Rightarrow$ (1) is contained in Theorem 3.

To simplify the notation, let $\Delta(t_1,t_2)$ stand for $\Delta_1(L)(t_1,t_2)$. By the second Torres condition $\Delta(t_1,1) = (t_1^{\ell} - 1/t_1 - 1)\Delta_1(L_1)(t_1)$ where $\ell = |\Delta_1(L)(1,1)|$ is the absolute value of the linking number, and analogously for $\Delta_1(1,t_2)$. Hence if $\Delta(-1,-1) = 0$, so $\Delta(1,1)$ is even, then $\Delta(-1,1) = \Delta(1,-1) = 0$. On expanding $\Delta(t_1,t_2)$ about $(-1,-1)$ as

$$\begin{aligned}\Delta(t_1,t_2) = {} & \Delta(-1,1) + a(t_1 + 1) + b(t_2 + 1) \\ & + (\text{terms of order} \geq 2 \text{ in } t_2 + 1 \text{ and } t_2 + 1),\end{aligned}$$

it follows that $\Delta(-1,-1) = 0$ implies that

$$0 = \Delta(1,-1) = 0 + a.2 + b.0 + (\text{terms of order} \geq 2 \text{ in } t_1 + 1 = 2 \text{ and } t_2 + 1 = 0)$$

and hence that a is even, and similarly b is even. Hence

$\Delta(1,1) = 0 + a.2 + b.2 + (\text{terms of order} \geq 2 \text{ in } t_1 + 1 = 2 \text{ and } t_2 + 1 = 2)$

is divisible by 4.//

Corollary. *If a 2-component link has Alexander polynomial zero, then it is a $\mathbb{Z}/2\mathbb{Z}$-boundary link.* //

It seems unlikely that in general $\eta(L) = \mu$ should imply that L be a $\mathbb{Z}/2\mathbb{Z}$-homology boundary link. In the case $\mu = 2$, the group $D = (\mathbb{Z}/2\mathbb{Z}) * (\mathbb{Z}/2\mathbb{Z})$ is fortuitiously metabelian, and thus amenable to discussion in terms of Alexander ideals.

An epimorphism $G \to D$ induces an epimorphism $G_2/G_3 \to D_2/D_3 \approx \mathbb{Z}/2\mathbb{Z}$ of the second stages of the lower central series of G. Hence if G is the group of a 2-component weak homology boundary link L, then the linking number of L is even [27]. (This also follows from Milnor's theorem.) Conversely the second Torres condition implies that a 2-component link L has even linking number if and only if $\Delta_1(L)(1,-1) = 0$. An argument analogous to that of Theorem 4 with $\mathbb{Z}^{\dagger} = \Lambda_2/(t_1 - 1, t_2 + 1)$ instead of $\widetilde{\mathbb{Z}}$ then shows that such a link is a weak $\mathbb{Z}/2\mathbb{Z}$-homology boundary link.

The condition on the linking number can be given a more geometric proof. Suppose that L is a 2-component link such that $L_1 = \partial M_1$ and $L_2 = \partial M_2$ where M_1 is a $\mathbb{Z}/p\mathbb{Z}$-manifold in S^3 and M_2 is a $\mathbb{Z}/q\mathbb{Z}$-manifold in $S^3 - M_1$. Then L_1 is homologous to $p.\beta(M_1)$ in $S^3 - M_2$ while L_2 is homologous to $q.\beta(M_2)$ in $S^3 - M_1$ (where $\beta(M_i)$ is the closed curve representing the Bockstein of the characteristic class of the singular manifold M_i [134]). Therefore the linking number of L_1 and L_2 is divisible by pq.

Definition A concordance $\mathcal{L} : \mu S^1 \times [0,1] \to S^3 \times [0,1]$ between links $L = \mathcal{L}|\mu S^1 \times \{0\}$ and $L' = \mathcal{L}|\mu S^1 \times \{1\}$ is a $\mathbb{Z}/2\mathbb{Z}$-boundary concordance if $\mathcal{L}$ extends to an embedding $\mathcal{P} : \coprod_{1 \leq i \leq \mu} W_i \to S^3 \times [0,1]$ of disjoint 3-manifolds which meet $\partial(S^3 \times [0,1])$ transversally and such that $\mathcal{P}(\partial W_i) = V_i \cup (\{i\} \times S^1 \times [0,1]) \cup V_i'$ where $V_i = \mathcal{P}(W_i) \cap (S^3 \times \{0\})$, $V_i' = \mathcal{P}(W_i) \cap (S^3 \times \{1\})$ and $\mathcal{P}|\{i\} \times S^1 \times [0,1] = \mathcal{L}|\{i\} \times S^1 \times [0,1]$. Furthermore $\mathcal{L}$ is a boundary concordance if all the 3-manifolds W_i may be assumed orientable.

If $\mathcal{L}$ is a $\mathbb{Z}/2\mathbb{Z}$-boundary concordance from L to L' then L and L' are each $\mathbb{Z}/2\mathbb{Z}$-boundary links; moreover $\mathcal{L}$ is a $\mathbb{Z}/2\mathbb{Z}$-boundary concordance if and only if there is a map of the group $\mathcal{G} = \pi_1(S^3 \times [0,1] - \text{im}\mathcal{L})$ to $\overset{\mu}{*}(\mathbb{Z}/2\mathbb{Z})$ carrying an ith meridian to the generator of the ith free factor.

Theorem 5 Let $\mathcal{L} : 2S^1 \times [0,1] \to S^3 \times [0,1]$ be a concordance from L to L', and suppose L is a $\mathbb{Z}/2\mathbb{Z}$-boundary link. Then $\mathcal{L}$ is a $\mathbb{Z}/2\mathbb{Z}$-boundary concordance, and hence L' is also a $\mathbb{Z}/2\mathbb{Z}$-boundary link.

Proof Let G and $\mathcal{G}$ denote the groups of L and $\mathcal{L}$ respectively. The natural map $j : G \to \mathcal{G}$ induces isomorphisms $j_n : G/G_n \to \mathcal{G}/\mathcal{G}_n$ on all the nilpotent quotients, by Stallings' theorem. The group $D = \mathbb{Z}/2\mathbb{Z} * \mathbb{Z}/2\mathbb{Z}$ has a presentation $\{u,v \mid u^2 = v^2 = 1\}$. Let $f : G \to D$ be a map carrying meridians x,y for L to (the images of) u and v respectively. Then f induces epimorphisms $f_n : G/G_n \to D/D_n$, and so there is a compatible family of epimorphisms $F_n = f_n \circ j_n^{-1} : \mathcal{G}/\mathcal{G}_n \to D/D_n$, and thus a map $F : \mathcal{G} \to \varprojlim (D/D_n)$. Now D_n is generated by $t^{2^{n-1}}$, where t is the image of uv in D, and $\bigcap_{n \geq 1} D_n = \{1\}$, so $\hat{D} = \varprojlim (D/D_n)$ is the completion of D with respect to the topology for which

$\{D_n\}$ is a neighbourhood basis at 1. This clearly induces the 2-adic topology on the infinite cyclic normal subgroup of D generated by t, and so there is an exact sequence

$$1 \to \mathbb{Z}_2 \to \hat{D} \to \mathbb{Z}/2\mathbb{Z} \to 1$$

where the conjugation action of $\mathbb{Z}/2\mathbb{Z}$ on the 2-adic integers $\mathbb{Z}_2$ is generated by multiplication by -1. Let $i : D \to \hat{D}$ denote the natural inclusion. The image of F is a finitely generated normal subgroup of $\hat{D}$ which contains i(D). Therefore $K = \mathbb{Z}_2 \cap \operatorname{im} F$ is a finitely generated torsion free abelian group which contains i(t). Let w generate the maximal cyclic subgroup of K containing i(t), and let K_o be a complementary summand to $\mathbb{Z}w$ in K. Then K_o is normal in im F, and $\operatorname{im} F/K_o$ is generated by the images of u and w. The map $\psi : D \to \operatorname{im} F/K_o$ sending u to $i(u).K_o$ and t to $w \circ K_o$ is an isomorphism, and so $\pi = \psi^{-1} \circ F : \mathcal{G} \to D$ is an epimorphism. Now $F(j(x)) = i(u)$ and $F(j(y)) = i(v) = i(u)i(t)$, so $\pi(j(x)) = u$ and $\pi(j(y)) = ut^k$ for some k in $\mathbb{Z}$. Since the abelianization of $\mathcal{G}$ is generated by the images of the meridians j(x) and j(y), k must be odd. Hence there is a commutative square

$$\begin{array}{ccc} \mathcal{G} & \xrightarrow{\alpha} & \mathbb{Z}^2 \\ \pi\downarrow & & \rho\downarrow \\ D & \xrightarrow{\beta} & (\mathbb{Z}/2\mathbb{Z})^2 \end{array}$$

where α maps j(x) to (1,0), j(y) to (0,1) and $\mathcal{G}'$ to (0,0), β maps u to (1,0), v to (0,1) and ρ is reduction modulo 2. Therefore $\ker\alpha/\ker\alpha \cap \ker\pi \approx \ker\beta$ is infinite cyclic, and so as in the discussion of the group Γ in Theorem 4, $\mathcal{G}/\ker\alpha \cap \ker\pi$ is generated by the images of j(x) and j(y). Therefore D is generated by the images of j(x) and j(y) and so k must be ± 1. The sign of k may be changed by composing π with conjugation by u. Thus it may be assumed that $k = +1$ and so $\pi(j(y)) = v$ and the theorem is proved. //

The last assertion of the theorem is also a consequence of the invariance of Murasugi nullity under arbitrary I-equivalence [88], and Theorem 4. The result of Kauffman and Taylor applies to links with any number of components; does Theorem 5 have a corresponding extension?

It is not true in general that a concordance between boundary links need be a boundary concordance. For let L be a 2-component 2-link with group G, and let B_1 and B_2 be disjoint 4-discs in S^4 which each meet each component of im L in a 2-disc and are such that the disc links $L|L^{-1}(B_1)$ and $L|L^{-1}(B_2)$ are trivial. Then $L|2S^2 - L^{-1}(\text{int}\, B_1 \cup \text{int}\, B_2)$ is a concordance from the trivial link to itself with group G. Now if G cannot map onto a free group of rank 2, as in the examples of Chapter II, no concordance with group G can be a boundary concordance.

In higher dimensions similar considerations apply. In particular every 2-component n-link is a $\mathbb{Z}/2\mathbb{Z}$-boundary n-link if $n \geq 2$, for its first Alexander ideal must vanish, as follows from Stallings' Theorem and Theorem V.2.

In other applications of nonorientable surfaces to classical knot theory, Clark has considered the minimum number of crosscaps of any nonorientable spanning surface for a knot as an invariant of the knot [28], while Gordon and Litherland have used nonorientable spanning surfaces in their algorithm for the Murasugi signature of a link [60]. (Cooper has remarked in conversation that the analogous Tristram-Viro signatures may be studied by means of $\mathbb{Z}/p\mathbb{Z}$-manifolds spanning the link.) A final question: is there a geometric proof that (2-component) slice links are $\mathbb{Z}/2\mathbb{Z}$-boundary links?

REFERENCES

We have included some references to the following topics not touched upon above:

Alexander polynomials of special classes of knots [12, 20, 64, 65a, 66, 109, 135, 150, 204];

Alexander polynomials in other geometric contexts [52a, 121];

Alternating links [36, 64, 101, 104, 136];

Computation [127a];

Conway's potential function [32, 86, 87];

Reidemeister torsion and K-theory [131, 196, 197, 199];

Representations [21, 65, 151, 152];

Symmetries (periodic knots) [21a, 65a, 66, 79, 122, 142];

Extensive lists of papers on various aspects of knot theory are given in the books of Birman [10], Crowell and Fox [43] and Neuwirth [146] and in the survey articles by Gordon [59], Kervaire and Weber [98], and Suzuki [186].

We use the following abbreviations below:

AS (-) = Annals of Mathematics Studies (-), Princeton University Press, Princeton;

EM (-) = Ergebnisse der Mathematik und ihrer Grenzgebiete (-), Springer-Verlag, Berlin-Heidelberg-New York;

GTM (-) = Graduate Text in Mathematics (-), Springer-Verlag;

LN (-) = Lecture Notes in Mathematics (-), Springer-Verlag;

and

MR (- : - -) = Mathematical Reviews, volume (-), review (- -), American Mathematical Society, Providence.

[1] A'Campo,N. Sur la monodromie des singularités isoleés d'hypersurfaces complexes
Invent. Math. 20 (1973), 147-169. MR 49:3201

[2] A'Campo, N. Le nombre de Lefschetz d'une monodromie
Indag. Math. 35 (1973), 113-118. MR 47:8903

[3] A'Campo, N. La fonction zêta d'une monodromie
Comment. Math. Helv. 50 (1975), 233-248. MR 51:8106

[4] Atiyah, M.F. and Macdonald, I.G. Introduction to Commutative Algebra
Addison-Wesley, Reading-Menlo Park-London-Don Mills, 1969.
MR 39:4129

[5] Auslander, M. and Buchsbaum, D.A. Invariant factors and two criteria for projectivity of modules
Trans. Amer. Math. Soc. 104 (1962), 516-522. MR 28:1215

[6] Bachmuth, S. Automorphisms of free metabelian groups
Trans. Amer. Math. Soc. 118 (1965), 93-104. MR 31:4831

[7] Bailey, J.L. Alexander Invariants of Links
Ph.D. thesis, University of British Columbia, 1977. (See also: A characterization of Alexander invariants of links Notices Amer. Math. Soc. Abstract 77T-G69).

[8] Barge, J. Dualité dans les revêtements galoisiens
Invent. Math. 58 (1980), 101-106. MR 81h:57007

[9] Baumslag, G. Groups with the same lower central series as a relatively free group. II. Properties
Trans. Amer. Math. Soc. 142 (1969), 507-538. MR 39:6959

[10] Birman, J. Braids, Links and Mapping Class Groups
AS 82, 1974. MR 51:11477

[11] Blanchfield, R.C. Intersection theory of manifolds with operators with applications to knot theory
Ann. of Math. 65 (1957), 340-356. MR 19 page 53

[12] Bleiler, S. Realizing concordant polynomials by prime knots
Pacific J. Math.

[13] Bourbaki, N. Commutative Algebra
Elements of Mathematics, Hermann-Addison Wesley, Paris-Reading, 1972. MR 50:12997

[14] Bourbaki, N. Algebra I
Elements of Mathematics, Hermann-Addison Wesley, Paris-Reading, 1974. MR 50:6689

[15] Brieskorn, E. Die monodromie der isolierten Singularitäten von Hyperflachen
Manuscripta Math. 2 (1970), 103-161. MR 42:2509

[16] Brody, E.J. The topological classification of lens spaces
Ann. of Math. 71 (1960), 163-184. MR 22:7125

[17] Brown, K.S. The derived subgroup of a free metabelian group
Arch. Math. (Basel) 32 (1979), 526-529.

[18] Buchsbaum, D.A. and Eisenbud, D. What makes a complex exact?
J. Algebra 25 (1973), 259-268. MR 47:3369

[19] Buchsbaum, D.A. and Eisenbud, D. What annihilates a module?
J. Algebra 47 (1977), 231-243. MR 57:16293

[20] Burde, G. Alexanderpolynome Neuwirthscher knoten
Topology 5 (1966), 321-330. MR 33:7998

[21] Burde, G. Darstellungen von Knotengruppen
Math. Ann. 173 (1967), 24-33. MR 35:3652

[21a] Burde, G. Über periodische knoten
Arch. Math. (Basel) 30 (1978), 487-492. MR 58:31051

[22] Cappell, S. A splitting theorem for manifolds
Invent. Math. 33 (1976), 69-170. MR 55:11274

[23] Cappell, S. and Shaneson, J.L. Link cobordism
Comment. Math. Helv. 55 (1980), 20-49.

[24] Casson, A. and Gordon, C.McA. On slice knots in dimension 3
in [128] Part 2, 39-53. MR 81g:57003

[25] Chen, K.T. Integration in free groups
Ann. of Math. 54 (1951), 147-162. MR 13 page 105

[26] Chen, K.T. Isotopy invariants of links
Ann. of Math. 56 (1952), 343-353. MR 14 page 193

[27] Chen, K.T. Commutator calculus and link invariants
Proc. Amer. Math. Soc. 3 (1954), 44-55 (erratum 993). MR 13 page 721

[28] Clark, B.E. Crosscaps and knots
Internat. J. Math. Math. Sci. 1 (1978), 113-123. MR 57:17620

[29] Cochran, D.S. Links with Alexander Polynomial Zero
Ph.D. thesis, Dartmouth College, 1970.

[30] Cochran, D.S. and Crowell, R.H. $H_2(G')$ for tamely embedded graphs
Quart. J. Math. (Oxford) 21 (1970), 25-27. MR 41:2660

[31] Cohen, M.M. A Course in Simple-Homotopy Theory
GTM 10, 1970. MR 50:14762

[32] Conway, J.H. An enumeration of knots and links, and some of their algebraic properties
in Computational Problems in Abstract Algebra (edited by J.C.Leech), Pergamon, Oxford, 1970, 329-358. MR 41:2661

[33] Cooper, D.
Ph.D. thesis, University of Warwick, 1980.

[34] Cooper, D. The universal abelian cover of a link
preprint, 1980.

[35] Cowsik, R. and Swarup, G.A. A remark on infinite cyclic covers
J. Pure Appl. Alg. 11 (1977), 131-138. MR 81e:57001

[36] Crowell, R.H. Genus of alternating link types
Ann. of Math. 69 (1959), 258-275. MR 20:6103b

[37] Crowell, R.H. Corresponding group and module sequences
Nagoya Math. J. 19 (1961), 27-40. MR 25:3977

[38] Crowell, R.H. The group G'/G" of a knot group G
Duke Math. J. 30 (1963), 349-354. MR 27:4226

[39] Crowell, R.H. The annihilator of a knot module
Proc. Amer. Math. Soc. 15 (1964), 696-700. MR 29:5241

[40] Crowell, R.H. Torsion in link modules
J. Math. Mech. 14 (1965), 289-298. (Note: this journal is now entitled "Indiana Math.J."). MR 30:4807

[41] Crowell, R.H. The derived module of a homomorphism
Adv. in Math. 6 (1971), 210-238. MR 43:2055

[42] Crowell, R.H. (letter to N.F.Smythe, May 1976).

[43] Crowell, R.H. and Fox, R.H. <u>Introduction to Knot Theory</u>
Ginn and Co., Boston 1963 Second Revised edition, GTM 57, 1977. MR 26:4348

[44] Crowell, R.H. and Strauss, D. On the elementary ideals of link modules
Trans. Amer. Math. Soc. 142 (1969), 93-109. MR 40:889

[45] Durfee, A.H. Fibred knots and algebraic singularities
Topology 13 (1974), 47-59. MR 49:1523

[46] Durfee, A.H. The characteristic polynomial of the monodromy
Pacific J. Math. 59 (1975), 21-26. MR 53:445

[47] Durfee, A.H. and Kauffman, L.R. Periodicity of branched cyclic covers
Math. Ann. 218 (1975), 157-174. MR 52:6731

[48] Fort, M.K. (Jr.) (editor) Topology of 3-Manifolds and Related Topics
Prentice-Hall, Englewood Cliffs, 1962. MR 25:4498

[49] Fox, R.H. A quick trip through knot theory
in [48], 120-167. MR 25:3522

[50] Fox, R.H. Some problems in knot theory
in [48], 168-176. MR 25:3523

[51] Fox, R.H. and Smythe, N.F. An ideal class invariant of knots
Proc. Amer. Math. Soc. 15 (1964), 707-709. MR 29:2798

[52] Fox, R.H. and Torres, G. Dual presentations of the group of a knot
Ann. of Math. 59 (1954), 211-218. MR 15 page 979

[52a] Franks, J.M. Knots, links and symbolic dynamics
Ann. of Math 113 (1981), 529-552.

[53] Gamst, J. Linearisierung von Gruppendaten mit Anwendungen auf Knotengruppen
Math. Z. 97 (1967), 291-302. MR 36:251

[54] Giffen, C.H. Link concordance implies link homotopy
Math. Scandinavica 45 (1979), 243-254.

[54a] Giffen, C.H. New results on link equivalence relations
preprint, 1976.

[55] Goldsmith, D. Symmetric fibred links
in Knots, Groups and 3-Manifolds (edited by L.P.Neuwirth), AS 84, 1975, 3-23. MR 52:1663

[56] Goldsmith, D. A linking invariant of classical link concordance
in [67], 135-170. MR 80h:57005

[57] Goldsmith, D. Concordance implies homotopy for classical links in M^3
Comment. Math. Helv. 54 (1979), 347-355. MR 80h:57006

[58] Gordon, C.McA. Knots whose branched cyclic coverings have periodic homology
Trans. Amer. Math. Soc. 168 (1972), 357-370. MR 45:4394

[59] Gordon, C.McA. Some aspects of classical knot theory
in [67], 1-60. MR 80f:57002

[60] Gordon, C.McA. and Litherland, R.A. On the signature of a link
Invent. Math. 47 (1978), 53-70. MR 58:18407

[61] Gutiérrez, M.A. Boundary links and an unlinking theorem
Trans. Amer. Math. Soc. 17 (1972), 491-499. MR 46:10000

[62] Gutiérrez, M.A. Polynomial invariants of boundary links
Rev. Colombiana Mat. 8 (1974), 97-109. MR 51:4211

[63] Gutiérrez, M.A. Concordance and homotopy. I. The fundamental group
Pacific J. Math. 83 (1979), 75-91.

[64] Hartley, R. On two-bridged knot polynomials
J. Austral. Math. Soc. 28 (1979), 241-249. MR 81a:57006

[65] Hartley, R. Metabelian representations of knot groups
Pacific J. Math. 82 (1979), 93-104. MR 81a:57007

[65a] Hartley, R. Invertible amphicheiral knots
Math. Ann. 252 (1980), 103-109.

[66] Hartley, R. and Kawauchi, A. Polynomials of amphicheiral knots
Math. Ann. 243 (1979), 63-70. MR 81c:57004

[67] Haussmann, J.-C. (editor) <u>Knot Theory</u>, Proceedings of a conference at Plans-sur-Bex, Switzerland, 1977
LN 685, 1978. MR 58:31084

[68] Hempel, J. Construction of orientable 3-manifolds
in [48], 207-212. MR 25:3538

[69] Hempel, J. 3-Manifolds
AS 86, 1976. MR 54:3702

[70] Hillman, J.A. A non-homology boundary link with Alexander polynomial zero
Bull. Austral. Math. Soc. 16 (1977), 229-236. MR 56:12300

[71] Hillman, J.A. High dimensional knot groups which are not two-knot groups
Bull. Austral. Math. Soc. 16 (1977), 449-462. MR 58:31098

[72] Hillman, J.A. Alexander ideals and Chen groups
Bull. Lond. Math. Soc. 10 (1978), 105-110. MR 57:17621

[73] Hillman, J.A. Longitudes of a link and principality of an Alexander ideal
Proc. Amer. Math. Soc. 72 (1978), 370-374. MR 80c:57006

[74] Hillman, J.A. Trivializing ribbon links by Kirby moves
Bull. Austral. Math. Soc. 21 (1980), 21-28. MR 81d:57005

[75] Hillman, J.A. Spanning links by non-orientable surfaces
Quart. J. Math. (Oxford) 31 (1980), 169-179. MR 81f:57002

[76] Hillman, J.A. A link with Alexander module free which is not an homology boundary link
J. Pure Appl. Alg. 20 (1981), 1-5.

[77] Hillman, J.A. The Torres conditions are insufficient
Math. Proc. Cambridge Philos. Soc. 89 (1981), 19-22.

[78] Hillman, J.A. Alexander polynomials, annihilator ideals and the Steinitz-Fox- Smythe invariant
Proc. London Math. Soc.

[79] Hillman, J.A. New proofs of two theorems on periodic knots
Arch. Math. (Basel)

[80] Hirschhorn, P.S. Link complements and coherent group rings
Illinois J. Math. 24 (1980), 159-163. MR 81a:57020

[81] Hitt, L.R. Examples of higher-dimensional slice knots which are not ribbon knots
Proc. Amer. Math. Soc. 77 (1979), 291-297. MR 80k: 57041

[82] Holmes, R. and Smythe, N.F. Algebraic invariants of isotopy of links
Amer. J. Math. 88 (1966), 646-654. MR 34:807

[83] Hopf, H. Fundamentalgruppe und zweite Bettische Gruppe
Comment. Math. Helv. 14 (1942), 257-309. MR 3 page 316

[84] Hosokawa, F. On ∇-polynomials of links
Osaka Math. J. 10 (1958), 273-282. MR 21:1606

[85] Husemoller, D. Fibre bundles
McGraw-Hill, New York-London-Sydney, 1966. Revised edition, with 3 appendices, GTM 20, 1975. MR 37:4821

[86] Kauffman, L.R. The Conway polynomial
Topology 20 (1981), 101-108.

[87] Kauffman, L.R. Combinatorics and knot theory
preprint, 1980.

[88] Kauffman, L.R. and Taylor, L.R. Signature of links
Trans. Amer. Math. Soc. 216 (1976), 351-365. MR 52:9210

[89] Kawauchi, A. On quadratic forms of 3-manifolds
Invent. Math. 43 (1977), 177-198. MR 58:7645

[90] Kawauchi, A. On the Alexander polynomials of cobordant links
Osaka J. Math. 15 (1978), 151-159. MR 58:7599

[91] Kawauchi, A. On links not cobordant to split links
Topology 19 (1980), 321-334.

[92] Kearton, C. Noninvertible knots of codimension 2
Proc. Amer. Math. Soc. 40 (1973), 274-276. MR 49:6217

[93] Kearton, C. Blanchfield duality and simple knots
Trans. Amer. Math. Soc. 202 (1975), 141-160. MR 50:11255

[94] Kearton, C. Cobordism of knots and Blanchfield duality
J. London Math. Soc. 10 (1975), 406-408. MR 52:6732

[95] Kearton, C. Signatures of knots and the free differential calculus
Quart. J. Math. (Oxford) 30 (1979), 157-182. MR 80h:57002

[96] Kervaire, M.A. Les noeuds de dimensions supérieures
Bull. Soc. Math. France 93 (1965), 225-271. MR 32:6479

[97] Kervaire, M.A. Knot cobordism in codimension 2
in Manifolds-Amsterdam 1970 (edited by N.H.Kuiper),
LN 197, 1971, 83-105. MR 44:1016

[98] Kervaire, M.A. and Weber, C. A survey of multidimensional knots
in [67], 61-134. MR 80f:57009

[99] Kidwell, M.E. Alexander polynomials of links of small order
Illinois J. Math. 22 (1978), 459-475. MR 58:12997

[100] Kidwell, M.E. On the Alexander polynomials of certain three-component links
Proc. Amer. Math. Soc. 71 (1978), 351-354. MR 58:2791

[101] Kidwell, M.E. On the Alexander polynomials of alternating 2-component links
Internat. J. Math. Math. Sci. 2 (1979), 229-237. MR 80g:57007

[102] Kinoshita, S.-I. On elementary ideals of polyhedra in the 3-sphere
Pacific J. Math. 42 (1972), 89-98. MR 47:1042

[103] Kinoshita, S.-I. On elementary ideals of θ-curves in the 3-sphere and 2-links in the 4-sphere
Pacific J. Math. 49 (1973), 127-134. MR 50:5775

[104] Kinoshita, S.-I. On the distribution of Alexander polynomials of alternating knots and links
Proc. Amer. Math. Soc. 79 (1980), 644-648. MR 81f:57003

[104a] Kinoshita, S.-I. The homology of a branched cyclic cover
preprint, 1980.

[105] Kirby, R. A calculus for framed links in S^3
Invent. Math. 45 (1978), 35-56. MR 57:7605

[106] Kirby, R. Problems in low dimensional topology
in [128] Part 2, 273-312. MR 80g:57002

[107] Kirby, R. and Melvin, P. Slice knots and Property R
Invent. Math. 45 (1978), 57-59. MR 57:7606

[108] Kojima, S. and Yamasaki, M. Some new invariants of links
Invent. Math. 54 (1979), 213-228. MR 81b:57004

[109] Kondo, H. Knots of unknotting number 1 and their Alexander polynomials
Osaka J. Math. 16 (1979), 551-559. MR 80g:57008

[110] Lambert, H.W. A 1-linked link whose longitudes lie in the second commutator subgroup
Trans. Amer. Math. Soc. 147 (1970), 261-269. MR 42:2470

[111] Laufer, H. Some numerical link invariants
Topology 10 (1971), 119-131. MR 42:8473

[112] Lê Dũng Trang Sur les noeuds algebriques
Compositio Math. 25 (1972), 281-321. MR 47:8541

[113] Levin, G. and Vasconcelos, W.V. Homological dimensions and Macaulay rings
Pacific J. Math. 25 (1968), 315-323. MR 37:6275

[114] Levine, J. Unknotting spheres in codimension two
Topology 4 (1965), 9-16. MR 31:4045

[115] Levine, J. A characterization of knot polynomials
Topology 4 (1965), 135-141. MR 31:5194

[116] Levine, J. Polynomial invariants of knots of codimension two
Ann. of Math. 84 (1966), 537-554. MR 34:808

[117] Levine, J. A method for generating link polynomials
Amer. J. Math. 89 (1967), 69-84. MR 36:7129

[118] Levine, J. Knot cobordism groups in codimension two
Comment. Math. Helv. 44 (1969), 229-244. MR 39:7618

[119] Levine, J. Knot modules I
Trans. Amer. Math. Soc. 229 (1977), 1-50. MR 57:1503

[120] Levine, J. Some results on higher dimensional knot groups
in [67], 243-269. MR80j:57021

[120a] Levine, J. Modules of 2-component links (See also Abstract 783-55-24, Amer. Math. Soc. (1981)).

[121] Libgober, A. Alexander polynomials of plane algebraic curves and cyclic multiple planes
preprint, 1980

[122] Lüdicke, U. Zyklische knoten
Arch. Math. (Basel) 32 (1979), 588-599. MR 81d:57007

[123] Magnus, W., Karrass, A. and Solitar, D. Combinatorial Group Theory
Interscience Publishers, New York-London-Sydney, 1966.
Second Revised edition, Dover Publications Inc., New York, 1976. MR 34:7617

[124] Masley, J. On euclidean rings of integers in cyclotomic number fields
J.Reine Angew. Math. 272 (1974), 45-48. MR 51:429

[125] Massey, W.S. Completion of link modules
Duke Math. J. 47 (1980), 399-420. MR 81g:57004

[126] Matumoto, T. On the signature invariants of a non-singular complex sesqui-linear form
J. Math. Soc. Japan 29 (1977), 67-71. MR 55:10386

[127] McMillan, D.R.(Jr.) Boundary-preserving mappings of 3-manifolds
in Topology of Manifolds (edited by J.C.Cantrell and C.H.Edwards, Jr.), Markham Publishing Company, Chicago, 1970, 161-175. MR 43:2723

[127a] Mehta, M.L. On a relation between torsion numbers and Alexander matrix of a knot
Bull. Soc. Math. France 108 (1980), 81-94.

[128] Milgram, R.J. (editor) Geometric Topology, Proceedings of Symposia in Pure Mathematics XXXII
American Mathematical Society, Providence, 1978. MR 80d:57001

[129] Milnor, J.W. Link groups
Ann. of Math. 59 (1954), 177-195. MR 17 page 70

[130] Milnor, J.W. Isotopy of links
in Algebraic Geometry and Topology, A symposium in honour of S. Lefshetz (edited by R.H.Fox, D.S.Spencer and W. Tucker), Princeton University Press, Princeton, 1957, 280-306. MR 19 page 1070

[131] Milnor, J.W. A duality theorem for Reidemeister torsion
Ann. of Math. 76 (1962), 137-147. MR 25:4526

[132] Milnor, J.W. Infinite cyclic covers
in Conference on the Topology of Manifolds (edited by J.G.Hocking), Prindle, Weber and Schmidt, Boston-London-Sydney, 1968, 115-133. MR 39:3497

[133] Milnor, J.W. Singularities of Complex Hypersurfaces
AS 61, 1968. MR 39:969

[134] Morgan, J.W. and Sullivan, D.P. The transversality characteristic class and linking cycles in surgery theory
Ann. of Math. 99 (1974), 463-544. MR 50:3240

[135] Morton, H. Infinitely many fibred knots having the same Alexander polynomial
Topology 17 (1978), 101-104. MR 81e:57007

[136] Murasugi, K. On alternating knots
Osaka Math. J. 12 (1960), 277-303. MR 25:563

[137] Murasugi, K. On the Minkowski unit of slice links
Trans. Amer. Math. Soc. 114 (1965), 377-383. MR 30:5309

[138] Murasugi, K. On a certain numerical invariant of link types
Trans. Amer. Math. Soc. 117 (1965), 387-422. MR 30:1506

[139] Murasugi, K. On Milnor's invariant for links
Trans. Amer. Math. Soc. 124 (1966), 94-110. MR 33:6611

[140] Murasugi, K. On Milnor's invariant for links. II The Chen group
Trans. Amer. Math. Soc. 148 (1970), 41-61. MR 41:4519

[141] Murasugi, K. On the signature of links
Topology 9 (1970), 283-298. MR 41:6198

[142] Murasugi, K. On periodic knots
Comment. Math. Helv. 46 (1971), 162-174. MR 45:1148

[142a] Murasugi, K. University of Toronto lecture notes, 1970.

[142b] Murasugi, K. (letter to J.A.Hillman, January 1980).

[143] Nakagawa, Y. On the Alexander polynomials of slice links
Osaka J. Math. 15 (1978), 161-182. MR 58:7600

[144] Nakanishi, Y. A surgical view of Alexander invariants of links
Math. Sem. Notes Kobe Univ. 8 (1980), 199-218.

[145] Neumann, H. <u>Varieties of Groups</u>
EM 37, 1967. MR 35:6734

[146] Neuwirth, L.P. <u>Knot Groups</u>
AS 56, 1965. MR 31:734

[146a] Neuwirth, L.P. The status of some problems related to knot groups
in <u>Topology Conference</u> (edited by R.F.Dickman, Jr. and P. Fletcher), LN 375, 1974, 209-230. MR 55:4129
(This review updates the problem list further).

[147] Papakyriakopolous, C.D. On Dehn's lemma and the asphericity of knots
Ann. of Math. 66 (1957), 1-26. MR 19 page 761

[148] Pardon, W. The exact sequence of a localization for Witt groups
in Algebraic K-theory (edited by M.R.Stein),
LN 551, 1976, 336-379. MR 58:5856

[149] Poenaru, V. A note on the generators for the fundamental group of the complement of a submanifold of codimension 2
Topology 10 (1971), 47-52. MR 42:6859

[150] Quach Thi Cam Van and Weber, C. Une famille infinie de noeuds fibrés cobordants à zéro et ayant même polynôme d'Alexander
Comment. Math. Helv. 54 (1979), 562-566. MR 81a:57012

[151] de Rham, G. Introduction aux polynômes d'un noeud
Enseignement Math. 13 (1967), 187-194. MR 39:2149

[152] Riley, R. Automorphisms of excellent link groups
preprint, 1977.

[153] Rolfsen, D. Isotopy of links in codimension two
J. Indian Math. Soc. 36 (1972), 263-278. MR 49:6248

[154] Rolfsen, D. Some counterexamples in link theory
Canad. J. Math. 26 (1974), 978-984. MR 52:9241

[155] Rolfsen, D. Localized Alexander invariants and isotopy of links
Ann. of Math. 101 (1975), 1-19. MR 52:1715

[156] Rolfsen, D. A surgical view of Alexander's polynomial
in Geometric Topology (edited by L.C.Glaser and T.B.Rushing), LN 438, 1975, 415-423. MR 52:11892

[157] Rolfsen, D. Knots and Links
Publish or Perish, Inc., Berkeley, 1976. MR 58:24236

[158] Rotman, J.J. An Introduction to Homological Algebra
Academic Press, New York-London, 1979. MR 80k:18001

[159] Rourke, C.P. and Sanderson, B.J. Introduction to Piecewise-Linear Topology
EM 69, 1972. MR 50:3236

[160] Rubinstein, J.H. Dehn's lemma and handle decompositions of some 4-manifolds
Pacific J. Math. 86(1980), 565-569.

[161] Rushing, T.B. Topological Embeddings
Academic Press, New York-London, 1973. MR 50:1247

[162] Sato, N.A. Algebraic invariants of Links of Codimension Two
Ph.D. thesis, Brandeis University, 1978.

[163] Sato, N.A. Algebraic invariants of boundary links
Trans. Amer. Math. Soc. 265(1981), 359-374.

[164] Sato, N.A. Free coverings and modules of boundary links
Trans. Amer. Math. Soc. 264(1981), 499-505.

[165] Sato, N.A. Alexander modules of sublinks and an invariant of classical link concordance
Illinois J. Math.

[166] Schubert, H. Die eindeutige Zerlegbarkeit eines Knotens in Primknoten
Sitzungsberichte Heidelberger Akad. Wiss. Math.-Natur Kl. 3 (1949), 57-104. MR 11 page 196

[167] Seifert, H. Über das Geschlecht von Knoten
Math. Ann. 110 (1934), 571-592.

[168] Serre, J.-P. Corps Locaux
Hermann, Paris, 1962. (English translation (by M.J.Greenberg) Local Fields GTM 67, 1979). MR 27:133

[169] Shaneson, J.L. Embeddings with codimension two of spheres in spheres and h-cobordisms of $S^1 \times S^3$
Bull. Amer. Math. Soc. 74 (1968), 972-974. MR 37:5887

[170] Shinohara, Y. On the signature of knots and links
Trans. Amer. Math. Soc. 156 (1971), 273-285. MR 43:1172

[171] Shinohara, Y. Higher dimensional knots in tubes
Trans. Amer. Math. Soc. 161 (1971), 35-49. MR 44:4763

[172] Shinohara, Y. and Sumners, D.W. Homology invariants of cyclic coverings with applications to links
Trans. Amer. Math. Soc. 163 (1972), 101-121. MR 44:2223

[173] Simon, J. Wirtinger approximations and the knot groups of F^n in S^{n+2}
Pacific J. Math. 90(1980), 177-190.

[174] Smythe, N.F. Boundary links
in <u>Topology Seminar</u>, <u>Wisconsin 1965</u> (edited by R.H.Bing), AS 60, 1966, 69-72. MR 34:1974

[175] Smythe, N.F. Isotopy invariants of links and the Alexander matrix
Amer. J. Math. 89 (1967), 693-704. MR 36:2139

[176] Smythe, N.F. n-linking and n-splitting
Amer. J. Math. 92 (1970), 272-282. MR 41:4520

[177] Spanier, E. <u>Algebraic Topology</u>
McGraw-Hill, New York, 1966. MR 35:1007

[178] Stallings, J.R. Homology and central series of groups
J. Algebra 2 (1965), 170-181. MR 31:232

[179] Steinitz, E. Rechteckige Systeme und Moduln in algebraische Zahlkörpern. I
Math. Ann. 71 (1912), 328-354.

[180] Stoltzfus, N. Unravelling the Integral Knot Concordance Group
Memoirs of the American Mathematical Society,
vol. 12, no. 192, 1977. MR 57:7616

[181] Sumners, D.W. A note on an example of Poenaru
Topology 11 (1972), 319-321. MR 45:4400

[182] Sumners, D.W. Polynomial invariants and the integral homology
of coverings of knots and links
Invent. Math. 15 (1972), 78-90 (erratum 17 (1973),
94). MR 45:1150

[183] Sumners, D.W. On the homology of finite cyclic coverings of
higher-dimensional links
Proc. Amer. Math. Soc. 46 (1974), 143-149. MR 50:3239

[184] Sumners, D.W. and Woods, J. The monodromy of reducible plane
curves
Invent. Math. 40 (1977), 107-141. MR 56:16643

[185] Suslin, A.A. (СУСЛИН, А.А. ПРОЕКТИВНЫЕ МОДУЛИ НАД КОЛЬЦОМ
МНОГОЧЛЕНОВ СВОБОДНЫ
ДОК. АН СССР, 229 (1976), 1063-1066.
(English translation: Projective modules over a polynomial
ring are free, Soviet Math. Dokl. 17 (1976),
1160-1164). MR 57:9685

[186] Suzuki, S. Knotting problems of 2-spheres in the 4-sphere
Math. Sem. Notes Kobe Univ. 4(1976), 241-371. MR 56:3848

[187] Swarup, G.A. Relative version of a theorem of Stallings
J. Pure Appl. Alg. 11 (1977), 75-82. MR 57:6206

[188] Terasaka, H. On null-equivalent knots
Osaka Math. J. 11 (1959), 95-113. MR 22:8511

[189] Torres, G. On the Alexander polynomial
Ann. of Math. 57 (1953), 57-89. MR 14 page 574

[190] Traldi, L. On the Determinantal Ideals of Link Modules and a Generalization of Torres' Second Relation
Ph.D. thesis, Yale University, 1980.

[191] Traldi, L. The $(\mu-1)^{st}$ elementary ideal of a link
Notices Amer. Math. Soc., Abstract 80-G92.

[192] Traldi, L. Linking numbers and the elementary ideals of a link
preprint, 1981. (See also Notices Amer. Math. Soc., Abstract 81T-G4).

[193] Tristram, A.G. Some cobordism invariants for links
Math. Proc. Cambridge Philos. Soc. 66 (1969), 251-264. MR 40:2104

[194] Trotter, H.F. Homology of group systems with applications to knot theory
Ann. of Math. 76 (1962), 464-498. MR 26:761

[195] Trotter, H.F. Torsion-free metabelian groups with infinite cyclic quotient groups
in The Theory of Groups (edited by M.F.Newman), LN 372, 1974, 655-666. MR 51:10482

[196] Turaev, V.G. (ТУРАЕВ, В.Г.) МНОГОЧЛЕН АЛЕКСАНДЕРА ТРЕХМЕРНОГО МНОГООБРАЗИЯ
МАТЕМ. СБ. 97(139)(1975), 341-359 (erratum 463).
(English translation: The Alexander polynomial of a three-dimensional manifold, Math. USSR Sb.26(1975), 313 - 329). MR 52:4306

[197] Turaev, V.G. (ТУРАЕВ, В.Г.) КРУЧЕНИЕ РЕЙДЕМЕЙСТЕРА И МНОГОЧЛЕН АЛЕКСАНДЕРА
МАТЕМ. СБ. 101 (143) (1976), 252-270.
(English translation: Reidemeister torsion and the Alexander polynomial, Math. USSR Sb. 30 (1976), 221-237). MR 55:6438

[198] Viro, O.Ya. (ВИРО, О.Я.) РАЗВЕТВЛЕННЫЕ НАКРЫТИЯ МНОГООБРАЗИЙ С КРАЕМ И ИНВАРИАНТЫ ЗАЦЕПЛЕНИЙ. I
ИЗВ. АН СССР, СЕР. МАТ., 37 (1973), 1241-1258.
(English translation: Branched coverings of manifolds with boundary, and invariants of links, Math. USSR-Izv. 7 (1973), 1239-1256). MR 51:6832

[199] Wagoner, J. K_2 and diffeomorphisms of two and three dimensional manifolds
in Geometric Topology (edited by J.C. Cantrell), Academic Press, New York-San Francisco-London, 1979, 557-577.

[200] Waldhausen, F. Whitehead groups of generalized free products
in Algebraic K-theory II (edited by H. Bass)
LN 342, 1973, 155-179. MR 51:6803

[201] Webb, P. (letter to J.A. Hillman, November 1980).

[202] Weber, C. Torsion dans les modules d'Alexander
in [67], 300-308. MR80h:57029

[203] Weber, C. Sur une formule de R.H. Fox concernant l'homologie d'une revêtement ramifié
Enseignement Math. 25 (1980), 261-272. MR 81d:57011

[204] Weber, C. Sur le module d'Alexander des noeuds satellites
preprint, 1980.

[205] Yajima, T. On a characterization of knot groups of some spheres in R^4
Osaka J. Math. 6 (1969), 435-446. MR 41:4522

[206] Yanagawa, T. On ribbon 2-knots. The 3-manifold bounded by the 2-knots
Osaka J. Math. 6 (1969), 447-464. MR 42:1101

[207] Yanagawa, T. On ribbon 2-knots. III. On the unknotting ribbon 2-knots in S^4
Osaka J. Math. 7 (1970), 165-172. MR 42:5249

[208] Osborne, R.P. and Zieschang, H. Primitives in the free group on two generators
Invent. Math. 63 (1981), 17-24.

[209] Sakuma, M. The homology groups of abelian coverings of links
Math. Sem. Notes Kobe Univ. 7 (1979), 515-530. MR 81g:57005

INDEX